T0134355

Sustainable Development in the 21st Century

Editor
Andreas Rechkemmer, Hamad Bin Khalifa University, Doha

Editorial Board
Kevin Collins, The Open University, Milton Keynes
Sven Bernhard Gareis, WWU Münster
Edgar Grande, WZB Berlin Social Science Center
Hartmut Ihne, Hochschule Bonn-Rhein-Sieg
Maria Ivanova, University of Massachusetts Boston
Uwe Schneidewind, Wuppertal Institute
Wilhelm Vossenkuhl, Ludwig Maximilians University of Munich

Volume 4

Ilka Roose

Flows of
Chilean Water Governance

Social Innovations in Defiance of Mistrust
and Fragmented Institutions

 **Nomos**

Erstgutachter: Prof. Dr. Dirk Messner
Zweitgutachter: Prof. Dr. Uwe Schneidewind

Tag der Disputation: 28.11.2019

Diese Arbeit wurde vom Fachbereich Gesellschaftswissenschaften der Universität Duisburg-Essen als Dissertation zur Erlangung des Doktorgrades (Dr. rer. pol.) genehmigt.

© Coverpicture: NASA Earth Observatory image by Joshua Stevens, using MODIS data from the Land Atmosphere Near real-time Capability for EOS (LANCE). Caption by Kathryn Hansen.

The Deutsche Nationalbibliothek lists this publication in the
Deutsche Nationalbibliografie; detailed bibliographic data
are available on the Internet at http://dnb.d-nb.de

a.t.: Duisburg/Essen, Univ., Diss., 2019

ISBN   978-3-8487-6930-8 (Print)
          978-3-7489-1019-0 (ePDF)

British Library Cataloguing-in-Publication Data
A catalogue record for this book is available from the British Library.

ISBN   978-3-8487-6930-8 (Print)
          978-3-7489-1019-0 (ePDF)

Library of Congress Cataloging-in-Publication Data
Roose, Ilka
Flows of Chilean Water Governance
Social Innovations in Defiance of Mistrust and Fragmented Institutions
Ilka Roose
259 pp.
Includes bibliographic references.

ISBN   978-3-8487-6930-8 (Print)
          978-3-7489-1019-0 (ePDF)

Onlineversion
Nomos eLibrary

1. Edition 2020
© Nomos Verlagsgesellschaft, Baden-Baden, Germany 2020. Printed and bound in Germany.

This work is subject to copyright. All rights reserved. No part of this publication may be reproduced or transmitted in any form or by any means, electronic or mechanical, including photocopying, recording, or any information storage or retrieval system, without prior permission in writing from the publishers. Under § 54 of the German Copyright Law where copies are made for other than private use a fee is payable to "Verwertungsgesellschaft Wort", Munich.

No responsibility for loss caused to any individual or organization acting on or refraining from action as a result of the material in this publication can be accepted by Nomos or the author.

Acknowledgments

This study was accepted as a dissertation in political sciences at the University of Duisburg-Essen (UDE) in the winter semester of 2019/20. I would like to thank the editors of the series »Sustainable Development in the 21st Century« for accepting the study and in particular Prof. Dr. Andreas Rechkemmer and Nomos for the good collaboration.

Throughout the writing of this dissertation I have received great support and assistance. Firstly, I would like to express my deep gratitude to Professor Dr. Dirk Messner and Professor Dr. Uwe Schneidewind, my research supervisors, for their patient guidance, encouragement and useful critique of this research work. Special thanks go out to Dr. Alex Godoy. It is owing to his extraordinary professional and personal qualities that I received the best mentoring I could wish for during my field stays in Chile. Furthermore, I deeply appreciate the professional and kind conversations with Dr. Annabelle Houdret.

This dissertation has benefited from the lively exchange of ideas with many friends and colleagues affiliated to the ARUS program at the UDE. I would like to thank all the members of the program for their fruitful advice and support. In particular, I would like to thank Dr. Elke Hochmuth, Professor Dr. Alexander J. Schmidt and PD Dr. Ani Melkonyan.

Moreover, I would like to thank the German Federal Enterprise for Academic Exchange Service (DAAD) and the Heinrich Böll Foundation for supporting this work with financial and ideological scholarships. Especially, I appreciate the enriching conversations with the head of the Heinrich Böll Office in Chile, Dr. Ingrid Wehr.

This study is based on many field visits that would not have been as fruitful without the generous support of many institutions and scholars in Chile. Firstly, I would like to thank the Universidad de Desarollo for its trust. Moreover, I would like to thank the Pontificia Univerdad Católica de Valparaíso and in particular the members of the Programa de Gestión Hídrica, Professor Eduardo Salgado, Nieggiorba Livellara, Marcela Sotomayor, Virginia Bravo, Carolina Olivares, Fernando Varas, and Viviana Álvarez, for their patience and assistance to my many questions and for making me feel like a member of their team. Likewise, I would like to thank the members of Ecosistemas, Juan Pablo Orrego, Mitzi Urtubia, Paulina Jineo and Patricia Salgado.

Acknowledgments

Without the warmest hospitality and friendship of the family Ardiles Vilches, I would not have been able to conduct this study. Moreover, I would also like to extend my thanks to Mario and Haydee, Terry and Eduardo for their generosity and encouragement.

Most importantly, I am very thankful to all my interview partners for opening their doors and minds to me and for sharing their precious time. I am deeply grateful for the trust they gave to me and my work.

Finally, I wish to thank my family and friends for their loving support, encouragement and patience throughout my study. I would like to especially mention Anika, Andrea and Luis. Furthermore, I am greatly indebted to countless persons who have inspired my professional and personal life throughout the last years and motivated me to do my best.

June 2020   *Ilka Roose*

Contents

Acknowledgments 5
Contents 7
List of Boxes, Figures and Tables 10
List of Abbreviations 13

Part 1:Introduction 15

1. Chile against the current: water governance facing climate change 15
1.1 Zooming into Chile: spatial patterns of water conflicts 19
1.2 The emblematic case of Petorca 22
2. Research question 23

Part 2:Conceptual framework and theoretical background 25

3. Definition of central terms 25
4. Conceptualizing and methods 27
5. An institutionalist perspective on social-ecological systems 33
5.1 Contributions of Ostrom's social-ecological system framework to this study 46
5.2 Actor-centered institutionalism in the light of water governance 49
    5.2.1 Actors 50
    5.2.2 Action orientation 52
    5.2.3 Action situation and scope of action 53
    5.2.4 Actors constellations and modes of action 53
5.3 Syntheses and the role of game theory and social innovation 55
6. Hypotheses 60

Part 3:Results 68

7. The social-ecological system of Petorca Province 68
7.1 Social, economic, and political setting 71
7.2 Related ecosystems 82
7.3 Resource system and resource unit 87
7.4 Governance systems 92

7.5   Actors and action orientation   103
        7.5.1   Actor group: civil society   106
                7.5.1.1 Local leaders   106
                7.5.1.2 Activists   113
                7.5.1.3 Citizens   121
        7.5.2   Actor group: governmental bodies and authorities   129
                7.5.2.1 Local authorities   129
                7.5.2.2 Regional authorities   136
                7.5.2.2 National authorities   142
        7.5.3   Actor group: economy   148
                7.5.3.1 Large-scale agriculture   148
                7.5.3.2 Water provider   159
        7.5.4   Others   161
                7.5.4.1 Padre Fernando   161
                7.5.4.2 Susana   164
                7.5.4.3 Maria   166
                7.5.4.4 Alfredo   169

Part 4: Discussion   172

8.   The focal action situation: flows of water governance in
        Petorca   172
8.1   Market-based institutional framework   173
8.2   Patterns of interaction orientation   181
        8.2.1 Individualistic behavior, competition and low
        cooperation   182
        8.2.2 Cooperative quality of collaborative projects in the field   185
        8.2.3 Social innovations in Petorca   191
8.3   Mechanisms of market-based water governance and low
        cooperation   194
        8.3.2 Low level of trust   195
        8.3.3 Transparency restricted by communication gaps and
                biases   196
        8.3.4 Lack of enforcement   198
        8.3.5 Inner, experienced heuristics, norms and rules   200
        8.3.6 Ambiguous impact of ecological factors   210
        8.3.7 Conection factors: institutional entropy and inequality   211
8.4   Leverage points for enhanced cooperation   216
        8.4.1 Fostering transparency with the help of trust-worthy
                facilitators   217
        8.4.2 Overcoming heterogeneity and rejection of politics by
                highlighting social and economic benefits   220

	8.4.3 Establishing information and fairness by self-organization and institutional bricolage	222
	8.4.4 Remaining challenges of social innovations	224
8.5	From here to where? Towards polycentric water governance on river basin scale	226
	8.5.1 Politics and polity suggestions	229
	8.5.2 Policy suggestions	234

Part 5: Conclusions	236

9.	Empirical findings	236
10.	Theoretical and practical significance of this study	239
11.	Limitations of this study and recommendations for future research	242

| References | 244 |

List of Boxes, Figures and Tables

Box 1:   Outstanding conflicts in Chile between 2015-2016   76

Box 2:   Description of coordinating body taking the example
            of Oficina de Asuntos Hidricos   232

Figure 1:   The names and number of all Chilean administrative
            regions and climatic regions. (Aitken et al. 2016)   20

Figure 2:   Actor mind map template.
            (Own illustration based on Scharpf 1997)   32

Figure 3:   The cooperation hexagon.
            (Adapted from Messner et al. 2013: 15)   58

Figure 4:   Synthesis of SESF and AI. (Adapted from McGinnis
            and Ostrom 2014: 4 modified by Roose based on
            Scharpf 2006)   60

Figure 5:   Upward spiral of social innovations. (Own illustration)   67

Figure 6:   Map of Petorca Province. (Adapted from DGA 2012
            in CNR and Universidad de Concepción 2016: 26
            modified by Roose)   69

Figure 7:   Income poverty in Petorca Province.
            (Data from BCN 2018a, b, c)   73

Figure 8:   Groundwater use Petorca River basin.
            (Own illustration derieved from data from DGA 2012
            in CNR and Universidad de Concepción 2016: 232)   90

Figure 9:   Groundwater use in La Ligua river basin
            (Own illustration derieved from data from DGA 2012
            in CNR and Universidad de Concepción 2016: 232)   90

Figure 10: Complex scenario of collective action in Petorca.
(Own illustration based on Ostrom 2005: 59).   195

Figure 11: Impact of social innovations (colored in red) on
complex scenario of collective action in Petorca.
(Own illustration based on Ostrom 2005: 59)   217

Figure 12: Polycentric water governance on river basin scale.
(Own illustration)   229

Table 1:   Types of actors in water governance. Categories of
AI shaded in grey. (Derieved from Hourdret and
Shabafrouz 2006, Scharpf 2006 and modified by Roose)   51

Table 2:   Categories and tiers. (Based on McGinnis and Ostrom
2014 and Roose)   70

Table 3:   Population growth, age and gender structure in Petorca,
La Ligua and Cabildo Districts.
(Data from BCN 2018a, b, c)   74

Table 4:   Reported crimes per year in Petorca, La Ligua and
Cabildo Districts. (Data from BCN 2018a, b, c)   77

Table 5:   Basic data about La Ligua River and Petorca River.
(Data from DGA 2018)   88

Table 6:   Water rights and water availability in Petorca Province.
(Data from MOP and DGA 2015)   91

Table 7:   Rule-making organizations. Public Sector.
(Aldunce et al. 2015 modified by Roose)   94

Table 8:   Rule-making organizations. Private Sector and Civil
Society. (Aldunce et al. 2015 modified by Roose)   95

Table 9:   Selected actor groups for interview analysis.
(Own research)   105

Table 10: Local leaders. (Own research)   107

Table 11: Activists. (Own research)   114

Table 12: Citizens. (Own research)   122

Table 13: Local authorities. (Own research)   129

Table 14:   Regional authorities. (Own research)   136

Table 15: National authorities. (Own research)   142

Table 16: Large-scale agriculture. (Own research)   149

Table 17: Water provider. (Own research)   159

Table 18: Padre Fernando. (Own research)   162

Table 19: Susana. (Own research)   164

Table 20: Maria. (Own research)   166

Table 21: Alfredo. (Own research)   169

Table 22: Orientation of Interaction between actor groups in
           Petorca Province. (Own research)   183

Table 23: Large-scale projects aiming to solve water scarcity in
           Petorca. (Own research)   187

Table 24: Social innovations (SI) in Petorca Province.
           (Based on TEPSIE (2014: 15) and
           Scheidewind et al. (1997) and own research)   191

List of Abbreviations

AC	Asociaciones de Canalistas (Canal Aassociation)
AI	Actor-centered institutionalism (Mayntz and Scharf 1995)
APR	*Comité o asociación de agua potable rural* (Potable water association or committee)
CASF	*Comunidad de Aguas superficiales* (Surface water community
CAST	*Comunidad de Aguas subterráneas* (Groundwater community)
CNR	*Comisión Nacional de Riego* (National Irrigation Commission)
CONADI	*Corporación Nacional de Desarrollo Indígena* (National Indigenous Development Corporation)
CORFO	*Corporación de Fomento de la Producción* (Chilean economic development agency)
CPR	Common-pool resources
DGA	*Direción Nacional de Aguas* (National Water Directive)
DOH	*Direción de Obras Hidraulicas* (Directive of Hydraulic Works)
FIA	*Fundación para la innovación agraria* (Foundation for Agricultural Innovation)
INDAP	*Instituto de Desarrollo Agropecuario* (Institute of Farming Development)
IAD	Institutional analysis and development framework (Ostrom 2005a)
JV	*Junta de Vigilancia*
LSAA	Large-scale agriculture association
M_1	Mechanisms underlying hipothesis 1
M_2	Mechanisms underlying hipothesis 2
MMA	*Ministerio del Medio Ambiente* (Ministry of Environment)
MINAGRI	*Ministerio de Agricultura* (Ministry of Agricultura)
MODATIMA	*Movimiento de Defensa por el acceso al Agua, la Tierra y la Protección del Medioambiente* (Movement in Defense of Access to Water, Land and Environmental Protection)
MOP	*Ministerio de Obras Públicas* (Ministry of Public Works)
OAH	*Oficina de Asuntos Hídricos*
ONEMI	*Oficina Nacional de Emergencia del Ministerio del Interior y Seguridad Pública* (National Emergeny Office of the Ministriy of Interior Affairs and Public Security)
PGH	*Programa de Gestión Hídrica*
PRODESAL	*Programa de Desarollo Local* (Local Development Program)
PUCV	*Pontificia Universidad Católica de Valparaíso*

SESF	Social-ecological system framework (Ostrom 2007, 2009)
SISS	*Superintendencia de Servicios Sanitarios* (Superintendence of Sanitary Services)
SI	Social innovation
SUBDERE	*Subsecretaría de Desarrollo Regional y Administrativo* (Undersecretariat for Regional and Administrative Development)
UTM	*Unidades tribunals mensuales* (monthly fiscal unit)
WUO	Water user organization

Part 1: Introduction

This dissertation investigates water governance and conflict. Based on the debate on current water governance, it studies the case of a water conflict in rural central Chile and discusses facilitators and pitfalls of the transition towards sustainable water governance (cf. Part 1). Institutions and human behavior have a major impact on the transition of socio-ecological systems. Therefore, I developed an analytical framework embedding institutionalist theory in a social-ecological system framework (SESF) to display in which way the institutional construct promotes or impedes the development towards sustainable water governance (cf. Parts 2 and 3). Firstly, my research argues that market-based paradigms have created an institutional system that deters trust between actors and therefore impedes cooperation and network benefits. Secondly, my work claims that water governance needs to promote local networks of social innovation and water user organizations to sustain ecosystem services by nurturing self-organized cooperation. At the same time, a governmental framework must be created that assures the basic right to water and enhances transparency to raise trust and experiences of fair reciprocity. Practical examples of cooperation deterring mechanisms of the current institutional system and leverage points towards cooperation used by social innovations are presented, paired with recommendations for water management policy (cf. Part 4). This dissertation provides a novel approach to water governance and institutionalism from a critical perspective, focusing on socio-ecological systems. Furthermore, it contributes to the rather new research on social innovation by putting it in the light of elaborate theoretical approaches. Moreover, this work contains research-based incentives for practical institutional changes towards sustainable water governance (cf. Part 5).

1. Chile against the current: water governance facing climate change

Water crises may be the most outstanding global crises in the future. According to the UNESCO world water report (UNESCO 2015), as soon as in 2030 global water demand will outreach our existing water resources by 40 % if we maintain current economic development. A major impact comes from the industry, which under existing conditions will increase its demand by 400 % (ibid.). According to the World Health Organization, half of the world´s population will be living in water-stressed areas by 2025 (WHO and UNICEF 2017). Business leaders recognize this development. According to the World Economic Forum (2018), water crises rank fifth place of global risks. However, this is not a problem of the future, water scarcity is affecting people already. In 2015 about 844 million people suffered from lack of access to drinking water and only 39 % had access to safely managed sanitation[1]. Poor sanitation can cause severe diseases and is estimated to cause more than 50 thousand diarrheal death each year (WHO and UNICEF 2017). The two prominent causes for water crises are closely linked to each other: (1) anthropologic climate change, and (2) population growth. Since the industrial revolution, the population has drastically increased leading to an almost eightfold increase of water demand in the last century (WBGU 2011). The international community has recognized those problems. After the United Nations (UN) Conference on Environment and Development in Rio de Janeiro 1992, globally recognized research, conventions and guidelines have framed political decisions and economic regulations normatively on an international level. The Intergovernmental Panel on Climate Change (IPPC) underlines that mitigation and adaptation to climate change are essential (IPPC 2007, 2014). In specific, it points out to the risks of water supply projected to increase with a global warming of 1,5°C respectively 2°C (IPPC 2018). In 2010 the UN recognized the human right to water and sanitation through the resolution 64/292[2] and in 2015 reconfirmed it in the General Assembly Resolution 7/169[3]. Moreover, they integrated access to water and sanitation for all into the Sustainable Development Goals (SDGs). SDG 6 aims inter alia for »*universal and equitable access to safe and affordable drinking water for all*« by 2030 (UN 2018). All these efforts show the dedication of the international community to solve water issues around the world. However, one question remains: Which local impact do those international agreements have (cf. Eid and Kranz 2014: 151; Whaley and Cleaver 2017)?

Anthropologic climate change and climate change impacts are related to social inequalities such as power structures, resource use or adaptation ca-

---

1   Meaning a toilet or improved latrine for one household that is connected to a safe
     disposal or treatment of excreta.
2   A/RES/64/292
3   A/C.3/70/L.55/Rev.1

pacity (Wuppertal Institut 2006; Brunnengräber 2011). Consequently, so-cial, political and geographical differences can lead to environmental con-flicts (Saretzki 2010). Therefore, several scholars have stressed the im-portance of society and the role of governance in environmental conflicts and climate change adaptation (cf. Paavola et al. 2006: 263; Baasch et al. 2012; Hyde 2014: 22; Heinrichs and Grunenberg: 2009: 14) As early as in the second UNESCO World Water Report (UNESCO 2006), the authors have pointed out that the water crisis is mainly a problem of water govern-ance. They highlight that water governance includes broad parts of society and politics outside the water sector (ibid: 47). Globally, the development of different water governance trends, which follow general development trends, can be observed throughout the last decades. Starting from strong state approach at the beginning of the second half of the 20th century, the states then rolled back and gave space for privatization during the 80s, 90s and onwards. This privatization mainly refers to privatization of water-services, such as supply or wastewater treatment (ibid.). The idea of mar-ket introduction was supposed to ensure the most profitable and efficient use of the resource aiming at its protection (UNCED 1993, Briscoe 1996). However, many scholars have criticized this »*open market's win-or-lose battle*« (Boelens 2008: 4) and warned about its dangerous impacts on community and its right to water. The expert on political ecology of water in Latin America, Rutgers Boelens and his colleagues draw attention to the social-environmental complexity of water governance by stating that water many times seems to flow in the direction of power, accumulating in the hands of a few dominant users who would tend to be more interested in turning it into an immediate benefit, instead of considering the conse-quences on the environment and health in the long term. Therefore, they argue that the unjust distribution of water is manifested not only in terms of poverty, but also constitutes a serious threat to food security and envi-ronmental sustainability (Boelens et al 2011).

Currently, water governance is moving towards an integrated communi-ty-based approach (Pahl-Wostl 2015), and a global trend of re-communalization of water services can be observed worldwide (Lobina et al. 2014). However, the step stones for faultless water governance have not been found yet. Hence, current research asks: Which are the processes towards sustainable water governance (cf. Pahl-Wostl et al. 2010)?

Chile, however, is striving against this current. In the early 80s, during the dictatorship of Agosto Pinochet, the Andean country separated land rights from water rights and opened its water market. In Chile water pri-vatization was introduced during a wave of privatization in different public sectors following neoliberal paradigms set by its constitution from 1980 onwards (cf. Bauer 2015). After the military dictatorship, specifically during the 90s, Chile privatized almost all water supply companies. The desired economic welfare failed to benefit the greater part of the Chilean society. Monopoles on natural resources formed (Benedikter and Siepmann 2015), and currently, Chile has the worst GINI Index[4] compared to the southern cone (The World Bank 2015) and ranks worldwide on the 15th place (CIA 2013). Due to the privatized water rights and water management, the Chilean privatization goes deeper than the globally observed trend. It is fostered in the Chilean constitution, which sets up high barriers to follow the global shift towards an integrated governance approach. On the one hand, Chiles water and sanitation policy is seen as a positive example because it has managed to drastically improve water and sanitation supply (cf. Molinos-Senante 2018; UNESCO 2017; Saleth and Dinar 2005; Valdés-Pineda 2014; Naciones Unidas, Gobierno de Chile 2010). On the other hand, scholars claim that challenges of climate change impacts and rising of water conflicts are either caused or supported by the dysfunctions of the institutional system, which is too weak and fragmented (Bauer 2015; Retamal et al 2013). A study of the World Resources Institute states that Chile will be facing extremely high (more than 89 %) water stress by 2040 due to rising temperatures and shifting precipitation patterns (Maddocks et al. 2015). The society is increasingly reacting to these changes. Important steps were taken between 2007 and 2009 when members of different parts of society (environmental, organizations, farmers, indigenous people, churches, water users organization, etc.), came together to start water protection coordination on a national level (Larraín 2010: 28). In literature, the dominant proposals for the improvement of Chilean water management focus on strengthening public and user participation (Clarvis and Allan 2014: 87–88; Retamal et al. 2013: 13), building trust through transparency (Clarvis and Allan 2014: 88) and improving adaptive capacity by paying attention to the social-ecological framework (Clarvis and Allan 2014: 88; Retamal et al. 2013: 9). However, not all scholars agree with each other. While the Chilean environmentalist Sara Larraín seeks for nationalization of water and independence from international policy (Larraín 2010: 21), Chilean water law specialist Alejandro

---

4   The GINI Index measures the extent to which the income or consumer distribution between individuals or households of one economy differs from completely equal distribution (The World Bank 2015).

Vergara Blanco demands to maintain the core paradigms of the existing regulation but to improve the bureaucracy and management of water user organizations (Vergara Blanco 2014). Being a member of the *Organisation for Economic Co-operation and Development* (OECD), Chile is required to act to meet the organization's guidelines for water (OECD 2005, 2015), which state inter alia improvement of water quality monitoring strategies, prioritizing water for human consumption and sanitation, and respect of ecological and social extraction limits (OECD/ECLAC 2016: 78). The extensive research on institutional changes and their connection to socio-ecological systems is needed (cf. Clarvis and Allan 2014: 88). Retamal et al. (2013:12) state that such academic effort should not focus on the establishment of environmental and financial indicators for water issues only, but also generate stronger support of the social sciences to analyse the behaviour of the diverse actors and their motivation. Consequently, Chile's water governance is opening questions about how neoliberal paradigms of water privatization deal with climate change and social and environmental inequalities?

1.1 Zooming into Chile: spatial patterns of water conflicts

Chile is facing an increase of water conflicts throughout the country (cf. Bauer 2015). While these conflicts are multi-dimensional in terms of being political, social, environmental respectively economic, they also vary on a geographical scale. Considering the overall water average, Chile benefits from an above average water availability compared with the world average and with the international value considered for sustainable development (cf. Valdés-Pineda et al. 2014). However, the overall average is deceptive because locally water availability varies due to high differences in precipitation and temperature: the northern region is characterized by the dry Atacama Desert, the central region has a Mediterranean climate, but has been hit by a severe drought, and the south is a rather cold and rainy region (cf. Figure 1).

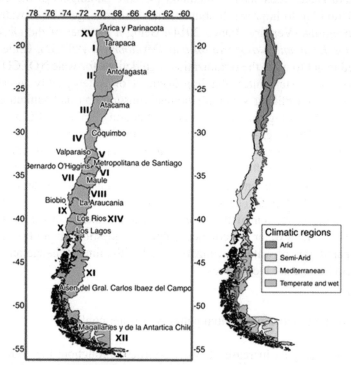

Figure 1: The names and number of all Chilean administrative regions and climatic regions. (Aitken et al. 2016)

Let me explain the differences by an example. While the northern region of Antofagasta only had a water availability of 53 (m3/inhabitant/year) in 2011, the southern region of Aysen had 2,993,535 (m3/inhabitant/year) (Valdés-Pineda et al. 2014).

Moreover, water use differs geographically. In Chile water is divided into consumptive water that does not return into the natural system (e.g. industrial water and potable water) and non-consumptive water that returns into the natural flow (90 % of non-consumptive water rights are hold by three hydroelectric companies, cf. Belmar et al. 2010). On national average, agriculture has the highest demand for consumptive water (77.8 %) and is followed by mining (7.2 %) (Valdés-Pineda et al. 2014). Furthermore, those demands also differ geographically. In the region of Antofagasta, for example, mining accounts for approximately 64.1 % of water use (Aitken et al. 2016). Thus, conflicts between actors with divergent interests over one catchment area differ geographically in relation to the present water use and water availability. According to the INDH (2012) most water conflicts in Chile are related to energy (approximately 43 %) and mining (approximately 34 %). As mining takes place mostly in the north, more water conflicts related to mining are found in this area. Water conflicts in central Chile concentrate on agriculture, and the southern part of Chile is predominantly confronted with conflict related to energy production (cf. Valdés-Pineda et al 2014; Rivera et al. 2016).

Moreover, in Chile, conflicts over water differ between the rural to the urban area. In both areas conflicts emerge due to competing demands of water users as explained above (normally, households vs industry; cf. Bauer 2015). However, urban and rural areas differ also in water supply and sanitation coverage. While the urban area has reached coverage for drinking water supply, wastewater collection and wastewater treatment of 99.9 %, 96.8 % and 99.8 %, respectively; in rural areas only 53 % of drinking water coverage has been reached (Fuster and Donoso 2018). This is also dependent on the fact that water supply and sanitation companies have been privatized. Mostly, they are in charge of urban areas. In rural areas, local associations of water supply (APR) have been established. Regulations for water provision are different in urban and rural areas. The most significant contrast is the lack of guarantee of good quality drinking water for rural citizens; here water supply companies only consult local people on questions about water provision. Usually, this role is delegated to local APRs (Donoso and Vicuña 2016). However, those APR often operate informally (about 13 % act without an authorization for sanitation, Fuster and Donoso 2018: 162). This means that those APRs mostly do not expect support from water supply companies. Additionally, only 11 % of APRs declare owning sewage systems (ibid.)

Furthermore, the human right to water for domestic consumption is not prioritized in the Chilean water regulations (cf. Donoso and Vicuña 2016: 8; Rivera et al. 2016: 40-41). As a consequence, in times of water scarcity in rural areas, residents must be supplied by water trucks (Donoso and Vicuña 2016: 35) while large scale agricultural farming keeps irrigating. The problems of water supply by trucks are manifold. First, this method does not allow for a continuity as the potable resources are limited, which – of course – also directly influences quantity. Secondly, water quality deteriorates significantly if water is provided in that manner. Additionally, financial costs (which are carried by the municipalities) of this solution are high (Donoso and Vicuña 2016: 33). Although it is considered as a temporal solution, in some rural areas water trucks have been supplying residents with drinking water for several years (cf. Gobernación 2016).

1.2 The emblematic case of Petorca

The expert on the Chilean water rights system, Carl J. Bauer, classifies conflicts between small- and large-scale farmers as the »worst« and refers specifically to the Petorca Province in the central-north of Chile (Bauer 2015: 158). This province of approximately 70 thousand inhabitants is divided into five different districts. The two river basins: La Ligua River and Petorca River are located in the districts of Petorca, La Ligua and Cabildo. However, both rivers have been declared to be dried out. Even before 2010, the area faced severe drought (Garreaud et al. 2017) and it has ever since. The boom of the high water demanding avocado fruit in the second half of the 20th century led to a spreading of monocultures and increased the local water stress. The socio-ecological consequences of the water scarcity are manifold. Water rights have been over-granted and illegal water withdrawal is widely recognized as common use (cf. Bauer 2015). On the other hand, a great part of rural communities depends on water trucks. Official data show that in 2016 almost one third of the population in the La Ligua District was provided by water trucks (2.9 % in Cabildo and 17.6 % in Petorca, Gobernación 2016).

While according to a study of the CNR and Universidad de Concepcion (2016) large-scale agricultural entrepreneurs make up only for 7.7 % of all farmers, they own approximately 90 % of the cultivated land. Hence, they dominate the area and directly benefit; meanwhile many small-scale farmers have not been able to adapt. In 2012, the INDH (2012) reported about 7000 impoverished small-scale farmers, and the growing export rates of large-scale agricultural farms did not compensate their loss in terms of employments (1996-2009, ODEPA 2018) As an example, between 2009 and 2013 the Petorca District showed 80 % of job decreases in the agricultural sector (BCN 2018b). All this has led to a conflict between the civil society, activists, large-scale agricultural companies and public authorities. The conflict is still ongoing and repeatedly receives nationwide attention in the media (cf. Carmona-Lopez 2018). The sitation in Petorca requires an investigation into how the current water management system is coping with climate change impacts (here: drought), water overexploitation and conflict in one of Chile most prospering economic sector: agriculture. Therefore, it is emblematic for current and upcoming water crises and has been taking as a case for this study.

RESEARCH NEEDED

Future development of water availability in Chile is non-optimistic (Valdés-Pineda et al. 2014). As a response to that, scholars study Chilean water management and climate change and underline the need to focus on its actors and social impacts (cf. Rivera et al 2016). Retamal et al (2013:12) recognize that current politics is focusing on management of water user organization. However, they state that we know little about their members' perception of water governance (ibid.). Clarvis and Allan (2014: 88) believe that enhancing the interaction of different stakeholders of one catchment area could improve the possibilities for environmental protection. All these claims point to an integrated social and ecological system approach (cf. Garreaud et al. 2017). They act as a base for the research question that is displayed in the following chapter as well as for the conceptual framework of this study (cf. Part 2).

2. Research question

In view of anthropologic climate change impacts such as droughts, and deterioration of water quality, it is widely recognized that »*the water crisis is largely a crisis of governance*,« as stated at the Second World Water Report (UNESCO 2006: 1). In the recent history, the global trend of water governance has shifted from the more governmental approach of a strong state over privatization and decentralization towards more integrated participatory approaches. Chile does not seem to move forward to the latter as it has maintained and intensified water privatizations since the early 80s. However, its water management has been well recognized for leading to an almost full coverage of water provision in urban areas (Fuster and Donoso 2018). At the same time, Chile is known to be one of the countries worldwide which will suffer most from water scarcity problems (Maddocks et al. 2015). Even today, the number of water conflicts in Chile is increasing and the country has been criticized for its weak regulative system and compliance assurance (cf. chapter 1). My research is built up on this debate of water governance acknowledging that institutions and human behavior have a major impact on the transition of socio-ecological systems. Based on that perspective, the research question arises:

»*In which way does the institutional framework promote or impede sustainable water governance in conflictive catchment areas?* «

Global international and national resolutions and regulations are aimed at sustainable water governance. Nevertheless, this canon of regulations has not yet effectively reached the civilian population (Whaley and Cleaver 2017: 56–57). At the same time, confined creative strategies have emerged from local communities to tackle the consequences of water scarcity and water crises. Thus, the knowledge about the emergence and potency of such dynamics could help decrease the vulnerability of societies, raise their adaptation capability and close the gap between (inter)national and local water governance. Petorca - a province north of Santiago de Chile – struggles with a water conflict. The main actors are small-scale farmers, citizens, politicians and large-scale farming companies of the region. Climate change, monoculture farming as a developing strategy, water privatization and missing regulations are known to be the main reasons for the drought (INDH 2012: 142).

The case of Petorca exposes the insufficiencies of (political) institutions, which are unable to provide an adequate solution to the problem. At the same time, self-organized initiatives have emerged claiming the increase of a sustainable and democratic change in water governance. I take Petorca as a reasonable example to answer the research question by discussing in how far the current institutional system enables sustainable water governance. Moreover, I demonstrate whether new local creative strategies (here called social innovations) can provide new procedures as solutions to the water conflict.

Part 2: Conceptual framework and theoretical background

Globally increasing water crises put pressure on the actors involved to develop appropriate water management strategies. In this context, research on water governance highlights that social and ecological systems need to be recognized as integrated systems that impact and depend on each other. On the one hand, Chile is known to have a prospering economy with almost complete water provision coverage. On the other hand, Chile suffers diverse kinds of water stress and conflicts. Based on that, this research aims to find pathways for a sustainable water governance taking the example of the Chilean central-north Petorca Province (cf. Part 1). Part 2 displays the conceptual framework and methods used in this study as well as the underlying theoretical background. Therefore, this part builds the ground on which the quantitative and qualitative data, which are presented in Part 3, have been collected and analyzed. The following Part 4 combines the insights of Part 2 and Part 3. It discusses the results with regard to the theoretical background. Moreover, it proposes a purposeful design for an institutional framework called polycentric water governance on river basin scale that aims at sustainability (cf. chapter 7.5). The final conclusions on the study's theoretical and practical implications and limitations of this study and recommendations for future research are displayed in Part 5.

3. Definition of central terms

This paper targets the question: »in which way does the institutional framework promote or impede sustainable water governance in conflictive catchment areas?« This question introduces several terms that need to be conceptualized.

Firstly, the term *sustainable water governance* is specified. The term water governance used in this study is a describing a management system that goes further than the common government-led perspective by involving all actors and levels that influence the resource management and is based on the definition of the UNDP as it points out its complexity and interdependence of internal and external factors:

»Water governance is defined by the political, social, economic and administrative systems that are in place, and which directly or indirectly affect the use, development and management of water resources and the delivery of water service at different levels of society. Importantly, the water sector is a part of broader social, political and economic developments and is thus also affected by decisions outside of the water sector.« (UNDP 2014)

In this research, *sustainable water governance* is considered as aiming for adaptive strategies towards long-term environmental and social resilience. This study emphasizes the human dimension, that has been neglected by water management approaches in the past (cf. Pahl-Wostl 2015). Current research on water governance highlights the importance of actors' integration such as »[...] *participatory management and collaborative decision making [and] more attention to management of human behaviour through* »*soft*« *measures [...].*« (Pahl-Wostl 2015: 17). Additionally, in politics public participation and the importance of community integrating approaches have been emphasized, for example in the Agenda 21 and the introduction of community based management (CBM) (Whaley and Cleaver 2017: 56). Based on that, this study classifies water governance as sustainable if it makes use of the (known) strategies towards an integrated management system.

Secondly, the research question assumes a connection between the institutional framework and water governance, which has been ascertained by several researchers (cf. Pahl-Wostl 2015, Vergara Blanco 2014; Ostrom 1990; Houdret and Shabafrouz 2006). Annabelle Houdret and Miriam Shabafrouz (2006: 11), for instance, affirm that the institutional context has severe influence on conflict or cooperation in the context of water governance, and Gopalakrishnan (2005: 04) states that cultural aspects, such as informal institutions, can be a determining factor for the function or dysfunction of an institutional context. The definition of *institutions* in this study follows the concept of Actor-Centered Institutionalism (Mayntz and Scharpf 1995, cf. chapter 5.2). It distinguishes between actors and institutions. The latter are dependent or independent variables which enable collective action. However, they do not necessarily determine it (ibid.). Here are some examples: Concerning water governance, formal institutions could be official rules and laws about water regulation. Informal institutions could be established through customs and habits and exist in the context of negotiations and arrangements (Houdret and Shabafrouz 2006).

A third term to be defined for this study is *conflictive catchment area,* also referred to as *water conflict.* This paper starts from a broad definition of water conflicts. Water conflict is not defined by violence but rather as a dispute between two or more actors, respectively actor groups, with divergent interests over the same water body. Although ecological factors, such as water quantity or quality, may trigger the situation, the conflict is seen as a reflection of socio-ecological structures, and therefore its causes are not restrained to environmental issues (cf. Baechler 2002: 529, Houdret 2008: 99-104)

Furthermore, the research question addresses the impact of *social innovation* on water governance. The term social innovation is defined in a multitude of ways. For my work, I have formed a definition based on Pol and Ville (2009: 881), Murray et al.. (2010: 3), and Moulaert (2013), as they consider economic as well as socio-political innovation and stress its characteristics of being a potential institutional actuator of change. Therefore, I define new creative strategies as social innovation if they (a) improve quantity or quality of life (cf. Poll and Ville 2009: 881) by responding to social needs of resources or services (cf. Pradel Miquel et al. 2013: 155), (b) create new social relationships by development of trust and empowerment within marginalized populations (cf. Murray et al. 2010: 3; Pradel Miquel et. al. 2013: 155) and (c) transform (formal or informal) institutions by changing governance mechanisms towards social inclusiveness (cf. Murray et al. 2010; Pradel Miquel 2013: 155). Also, but not necessarily, (d) social innovations are locally embedded as they change social relationships which are mostly spatially connected (Dawson and Daniel 2010).

4. Conceptualizing and methods

The case of Petorca is a domestic water conflict over one catchment area (consisting of two river basins). It is classified as a collective good problem which consists of distribution and access problems (cf. chapter 6). By taking Petorca as a case, this paper is a hypothesis refining (and generating) case study that seeks to »*clarify the meaning of certain variables and the validity of empirical indicators, [and may] suggest alternative causal mechanisms and identify overlooked interaction effects*« (Vennesson 2008: 227-228). The in-depth case analysis allows for detailed description and understanding of the actor's perception. Hence, it captures the complexity and the context-bound factors of the stated research question. Furthermore, it is expected to contribute to theory development by discovering unidentified causal processes (cf. della Porta 2008: 211). Although an in-depth case study generates strong context bound data and, accordingly, its transferability is limited, this study follows criteria such as objectivity, validity and reliability to assure transparency, controllability and, if possible, replicability. The time frame chosen for this study corresponds to the empirical research stay in the case study area and the data available. The empirical research of the present study started in 2015 and was conducted until 2017. However, the available data vary and cover predominantly the timeframe from 2013 to 2015. Therefore, the study focusses on the timeframe from 2013 to 2017.

To define a suitable analytical framework that aims to answer the research question about Petorca, literature on the theories of institutionalism, commons governance and social-ecological systems were reviewed. After doing so, a synthesis of different approaches appealed to be most reasonable for the present study. Therefore, it uses the social-ecological system framework (SESF) by Ostrom (2007, 2009) as an explanatory framework and combines it with the actor-centered institutionalism (AI) by Mayntz and Scharpf (1995) for further causal reconstruction. In addition, insights of social innovation approaches have been used to add factors to the list of indicators set by SESF and AI. Chapter 5 displays both concepts and their synthesis in detail.

Furthermore, this research applies process-tracing to find mechanisms between institutional framework and sustainable water governance. Process-tracing facilitates the linking of a specific cause X to a specific outcome Y by exploring the underlying causal and descriptive mechanisms M (cf. Vennesson 2008; Mahoney 2012). Since this study emphasizes human behavior in water governance, interaction in terms of cooperation is central for answering the research question. Accordingly, the hypotheses in this research state:

- The market based institutional framework restricts cooperation in Petorca. (HP1)
- Social innovations overcome the market based institutional framework and enhance cooperation. (HP2).

Applying process-tracing, this study regards market based institutional framework as the cause X_1 for restricted cooperation Y_1. Moreover, it states that social innovation X_2 enhances cooperation Y_2. The underlying mechanisms between X and Y are called M_1 respectively M_2. This study uses the indicators of a SESF as well as AI as an analytical base to test the existence of X_1 and X_2 as well as Y_1 and Y_2, and to explore the connecting mechanisms M_1 and M_2. Scharpf (2006) suggests using game theory for modeling and understanding the investigated political problem. Although this study bases its argumentation on insights of game theory in the process of hypothesis building, it abandons archetypal constellations of game theory as it would reduce the complexity of the presented case. The hypotheses are presented in detail in chapter 6.

The findings of the applications of the SESF and the AI are displayed in Part 3: Results. The chapters of this part are arranged according to the categories of the SESF. Chapters 7. to 7.4 concentrate on the SESF following a descriptive order from a macro to a micro level. They first display data about the social, economic and political settings in general. Then, they present relevant data about the resource system. Chapter 7.4 focuses on the details of the water governance system in the research area of Petorca. The following chapter 7.5 builds up the core analysis. It consists of the synthesis between SESF and AI by displaying the perceptions of different actor groups.

In the following Part 4: Discussion, the data displayed in the previous chapters are used to test the hypotheses stated above. Chapter 8.1 focuses on testing the existence of X_1: market based institutional system. Chapter 8.2 proves the outcome Y_1: low cooperation level. Testing the cooperation level in 8.2 shows that especially small-scale projects reach a higher cooperation level. These small-scale projects fit the definition of social innovation and therefore prove the existence of X_2 (social innovation) and Y_2 (enhanced cooperation). The following chapter 8.3 filters the linking mechanisms M_1 that restrict cooperation and connect X_1 to Y_1. Chapter 8.4 then takes a closer look at the social innovations discovered in chapter 8.3. By filtering the mechanisms M_2 between social innovations and enhanced cooperation, it shows in which way M_2 copes with respectively overcomes the restricting mechanisms M_1. The last chapter 8.5 discusses in how far such social innovations have a potentially repercussive effect on the institutional system and may alter it towards a more sustainable water governance. Furthermore, it proposes an institutional design that aims for sustainability by polycentric water governance on river basin scale.

METHODS

Indicators of a SESF are used as a base for the data collection. First, the indicators relevant to the present research case (cf. chapter 7) were filtered. This list is divided into social, economic and political setting; related ecosystems; resource system; governance system and actors and action orientation. Then, the approach of del Mar Delgado-Serrano and Ramos (2015) was followed and each indicator regarding its scale (local, regional, national, international), research tool (observation, interview, media analysis, database), data format (numerical, narrative description) was categorized. Moreover, practical problems that might interfere in the data collection were added. In compliance with the research tool, I used different methods to collect data. The data of social, economic and political setting, related ecosystems, resource system and governance system were mostly collected via secondary sources such as governmental databases and legal documents. The data of the field actors and action orientation build the core part of this study and were collected by qualitative empirical research. Since a mix of quantitative and qualitative data was used, it could be argued that this study applies mix-methods research. However, the data analysis and discussion concentrate on qualitative data. Therefore, this study is categorized as qualitative. The quantitative data serve for completion of the descriptive work to capture the case's complexity.

In the context of the qualitative empirical research 53 interviews and multiple field observations were conducted (community and congress meetings, activists' meetings etc.) during the first field stay in 2016 and five further interviews during the field stay in 2017. The interviews followed a semi-structured interview guide which was based on the categories of SESF combined with AI. The interview guide was tested with regard to its feasibility (e.g. comprehensibility to the interview partner) and was reworked several times. The interviews were mostly held in a calm atmosphere at the interview partner's houses or offices. The interviews were kept open for the partners to add information and topics. Their length ranged from 30 minutes to two hours while the majority was held in about 60 minutes. Most interviews were recorded on audio tape while some were documented as written protocols. The interview guide was adjusted to the interview partners with regard to their actor group. For example, farmers were asked about the size of their cultivated land, their crops and farming strategies. Another example is the adjustment of language used. It showed that the term water conflict referring to the core situation of »*dispute between actor groups with divergent interests over the same water body*«, had to be adjusted in the interviews. As the use of the term water conflict seemed to be partisan to some actor groups, it was replaced by drought. However, this term was perceived as partisan to other actor groups as it underlines a natural cause. Therefore, the situation was finally referred to as water scarcity.[5]

---

5   A template of a typical interview guide and its connection to the (sub-)tiers of SESF can be found in the online annex (www.nomos-shop.de/isbn/978-3-8487-6930-8).

For each actor group and subgroup, I chose three to four interviews out of the collected data. In total, 36 out of 53 interviews were selected for transcription and in-depth analysis.[6] In order to obtain a wider range of perspective, I selected interviews of partners of different age, sex and educational background, when possible. My memories of the conducted interviews and the coding described in the following led to my decision whether further interviews should be added to a subgroup or whether the information was satisfactory. After doing a commented transcription of the selected audio material by documenting certain distinctive features of the spoken words like pauses or hesitation but not every variation of phonology (Höld 2009: 661), I used a mix of inductive and deductive categorizations based on the method of qualitative content analysis by Mayring (2010). For this, I coded the interviews with special attention to statements referring to cognitive aspects, motivational aspects and preferences as well as interaction orientation. According to Scharpf (2006), these three constitute the action orientation according to the social unit of the interview partners (see chapter 5.2.2). The following Figure 2 gives an overview of the deductive structure of the interview analysis process of code building. The codes are named in accordance with the spoken word and then clustered according to the categories of cognitive aspects, motivational aspects and preferences, and orientation of interactions (see Figure 2). To further cluster the codes, I used the sub-tiers developed through the synthesis of SESF by Ostrom (2007, 2009) and AI by Mayntz and Scharpf (1995). The coding was supported by using the qualitative data analysis software MAXQDA 12. After each interview coding, I compared the codes within the same interview and with previous interviews to avoid duplication. In the case of two very similar codes, I reduced the list to only one[7] Taking this as a base, I developed different Actor Group Mind Maps, that help exploring the actor behavior.[8]

---

6   A complete list of selected interview partners can be found in the online annex (www.nomos-shop.de/isbn/978-3-8487-6930-8).

7   A list of codes can be found in the online annex (www.nomos-shop.de/isbn/978-3-8487-6930-8).

8   An example of an actor group mind map can be found in the online annex (www.nomos-shop.de/isbn/978-3-8487-6930-8).

In 2017, five additional interviews were conducted with people related to identified social innovations. The semi-structured interview guide was based on the questions to analyze social innovations proposed by González et al. (2010: 57–64) and Pradel Miquel et al. (2013: 159-160).[9]

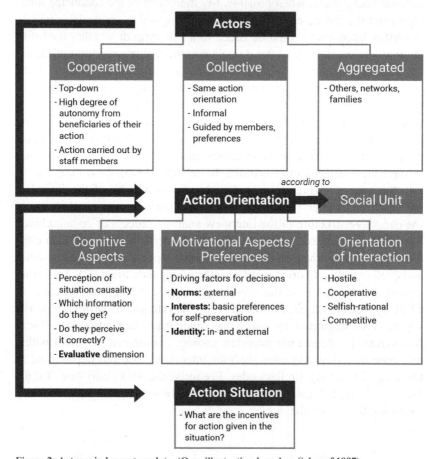

Figure 2: Actor mind map template. (Own illustration based on Scharpf 1997)

9   A template of a typical interview guide for social innovations can be found in the online annex (www.nomos-shop.de/isbn/978-3-8487-6930-8).

In addition to the method of semi-structured interviews, open observation was used to enhance qualitative data of cognitive and motivational aspects and interactions of actor groups. In the framework of this study, I attended different meetings (eight meetings with WUOs, one meeting with governmental program for small-scale farmers (PRODESAL), one meeting with the Governor and civil society, two congress meetings, two activists' meetings, one meeting with the National Water Directive (DGA)). In most of them, I introduced myself in the beginning, but I stayed passive during the rest of the meeting to keep my impact on the actors' behavior as small as possible. The observations were documented by protocols and served to complete the qualitative data of the perception analysis in chapter 7.5.

5. An institutionalist perspective on social-ecological systems

This study embeds an institutional approach into a social-ecological system context to tackle the research question of in which way institutional framework is promoting or impeding sustainable water governance in conflictive catchment areas (cf. chapter 4). This analytical framework has been developed based on a review of existing theoretical approaches. Its derivation and design are described in the following paragraphs.

Why connecting social and ecological systems? A challenge in analysing common resource problems such as water scarcity or water conflict is coping with complexity and integrating all relevant internal and external factors. Recent incidents like natural disaster and climate change impacts increased the concerns about the vulnerability of our socioeconomic system. This has led to a common wish of a sustainable development, which would create and maintain a prosperous social, economic and ecological system (Folke et al. 2002: 437). Ecological system components are put under high pressure by human interventions. As humans dominate the earth's ecosystem, ecologists argue if they should be included in their ecosystem models. By that way, they argue, an integrated approach will be reached that is more realistic and useful for implementation in real sociopolitical circumstances (Evans 2011: 225; Fiksel 2006: 14). In that sense, human and natural systems will no longer be treated as independent investigation fields. Additionally, facing current events, such as earthquakes, tsunamis or storms, scientists must admit that the ecosystem's responses to the human impact are neither linear nor predictable nor controllable (Folke et al. 2002: 437). A plausible example of this interplay is given by the Chilean environmentalist Larraín (2010: 22) as she emphasizes how hydro power plants in Chile are discussed as a source of clean energy while disregarding costs of the social and environmental hazards that they evoke, such as loss of biodiversity or removal of indigenous people. The approach of social-ecological system frameworks responds to this issue. In this context space is regarded as an integrated social and ecological system and therefore, the complexity and interdependency of biophysical and social factors is acknowledged (Evans 2011). This approach accepts the unpredictability of the complexity which occurs within this coherent system of biophysical and social factors. Management decisions made on a social issue of a socio-ecological system may have impacts on its ecological diversity. This framework forces governance to work interdisciplinarily and to be aware of the unknown. Consequently, it helps to implement ecological principles into socioeconomic structures in order to strengthen the resilience of the whole system.

RESILIENCE IN THE LIGHT OF SOCIO-ECOLOGICAL SYSTEMS
Resilience plays a central role in socio-ecological approaches. The term resilience has been used in many different sciences such as engineering, psychology or economy (cf. Timmerman 1981, Fiksel 2006). Social-ecological research regards resilience as an analytical category, which describes the breakdown susceptibility of a system, facing threats or shocks. According to Fiksel (2006) resilience is the »*capacity of a system to tolerate disturbances, while retaining its structure and function*«. Another more detailed definition of resilience in ecological sciences reads as follows: Resilience is the »*capacity of a system to absorb disturbance and reorganize while undergoing change so as to still retain essentially the same function, structure, identity, and feedbacks*« (Walker et al. 2004: 2) These definitions show that resilience is about equilibrium. Though, it must be recognized, that equilibrium could be meant as a steady-state situation, like it is termed as engineering resilience by Holling and Gunderson (2002: 27), or as a dynamic equilibrium. In the second case, Holing and Gunderson term it as ecosystem resilience. Resilience in ecosystem theories is not a static feature of a system but grows and shrinks in the different phases a system can face. Therefore, to implement resilience in management or policy approaches towards sustainability, it is important to have in mind the existence of more than one state of equilibrium. The term resilience used in this study, follows the concept of resilience in ecosystem theories as the concept of engineering resilience strengthens the idea

of predictable consequences and does not recognize appropriate the complexity of the system (Holling and Gunderson 2002: 28).

Since, the term resilience is not easy to conceive. Often it is set as the opposite to vulnerability or as the same as adaption. To get a more evident comprehension Walker et al. 2004 described resilience by four dimensions: the probability of loss or threat to the system - termed as precariousness, the sensibility for threat - termed as latitude, the inertia of a system, here called resistance, and the dependence of one system to another named panarchy. Precariousness describes the probability of a threat and therefore its urgency. It shows »*how close the current state of the system is to limit or »threshold«* (Walker et al. 2004). In this case resilience of a system could be increased by shifting tipping points. Latitude tells about the extent of damage if a shock or threat occurs. To strengthen resilience, the system could lower the damage in order to survive. In the case of climate change, adaption strategies would decrease the latitude of the (socio-) ecological system and therefore higher its resilience. Resistance is about the ability of a system to implement changes and to use the theoretical scope of action. Taking the example of climate change, resistance is the ability to implement mitigation and adaptation strategies. Panarchy describes the dependence of the system to others. In the view of climate change, the interrelation between the systems of a rain forest and cities can be called panarchy. With these four dimensions, resilience can be defined easier, but it is important to notice that none of these concepts alone are equal to resilience.

LINKING ISNTITUTIONS AND SOCIAL-ECOLOGICAL SYSTEMS

Folke et al (2002) see institutions as the link between the social and the ecological system as they frame the collective use of natural resources. Following this, I concentrate on institutionalism to make an in-depth analysis of the actors and actors' orientation. The following paragraphs display different strains of institutionalism and their connection to socio-ecological systems. Furthermore, an explanation why I chose the SESF by Ostrom (2007, 2009; McGinnis & Ostrom 2014) to embed the analytical framework of AI developed by Mayntz and Scharpf (1995) will be given followed by an extended description of each approach and its synthesis.

WHAT DO WE KNOW ABOUT INSTITUTIONAL APPROACHES?

The institutional approach studies individual behaviors and institutional dynamics in common property contexts – such as water conflicts and conflicts about other natural resources – with the aim of generating a predictive theory of collective action on suitable common property management (Johnson 2004). In this chapter, I outline different strains of this perspective and its advantages and disadvantages regarding common resource analysis. In order to focus on water and water conflicts, I mostly draw on the insights given by the political scientist Annabelle Houdret and the professor for resource management Claudia Pahl-Wostl. The work of both focuses on water. The approach to explain political and societal phenomena by institutions is not a new practice as it has already been applied before – for example in the work of Rousseau. However, since the late 70's it has been resumed (Schulze 1997) and, consequently, become labeled *New Institutionalism* (Schulze 1997, March and Olsen 1984).

Since *New Institutionalism* emerged, several strains have developed that are based on different assumptions. To classify institutionalism into different schools of thoughts, there are several categorizations to be found. While Peter A. Hall and Rosemary C. Taylor (1996) outline three types of *New Institutionalism: (1) Historical Institutionalism* coined especially by North (1993), (2) *Rational Choice Institutionalism* and (3) *Sociological Institutionalism*, Janice C. Romzek (2009) and B. Guy Peters (2005) conclude that there are seven strains of *New Institutionalism* in total. They add four more types to the categories of Hall and Taylor (1996): (4) *Normative Approach* by March and Olsen (1986), (5) *Empirical Institutionalism*, (6) *International Institutionalism* and (7) *Network Institutionalism*. Renate Mayntz and Fritz W. Scharpf (1995) also define three categories but call them (1) *Economical Institutionalism*, (2) *Institutional Organizational Sociology* and (3) *Neo Institutionalism of Political Science*, whereas Holger Schulze (1997) divides the different strains in four categories: *(1) Rational Choice Approach, (2) Historical-Economical Approach, (3) Historical Sociological Approach, (4) Sociological / Organizational Approach*. The understanding of how institutions develop differs depending on the scholarly direction followed. For this reason, Schulze (1997: 8) defines the different strains by their comprehension of institutions – from institutions made by aggregated rational choices of individuals on the one side (in case of rational-choice approach), to an integrated institution building on the other side (in case of sociological-/organizational approaches).

Following the different categorizations pointed out above, there are three strains underlined by each scholar: *Rational Choice Approach, Historical Approach* and *Sociological Approach*. The further approaches can be defined as subcategories, that are assigned to one of these or in-between these three. Therefore, I will now give a short explanation of these strains and outline their significant features and their usefulness regarding the analysis of actors' behavior in water conflicts. The most remarkable split lies between the *Normative* and the *Rational Choice Institutionalism*. The main difference between these two can be described as followed: The rational choice theory states that institutions are made on the bases of utility maximization and provide incentives or disincentives for actors' behavior. Following this theory, actors are fully autonomous, utility-maximizing and – of course – fully rational (Romzek 2009: 25). Due to their aggregate decisions, institutions will be formed in order to fulfill their needs (Schulze 1997). Thus, institutions themselves are seen as a product of the individual's agreement to find the most productive way of acting out utility-maximizing (ib.). Therefore, they restrain opportunistic behavior and try to overcome practical dilemmas caused by collective action (Hall and Taylor 1996). It has been shown that this approach is quite useful if focusing »*primarily on the functions that (these) institutions perform and the benefits they provide.*« (Hall and Taylor 1996: 19–20). Nevertheless, this approach has been criticized on different levels, especially for missing out explanations for inefficiencies especially on water conflicts and for its abstractedness (Houdret 2010; Hall and Taylor 1996). The main criticism lies on the limited explanations of social behavior as rational choice approaches focus on maximizing profits only, and therefore are not sufficient for finding the roots and development reasons of resource conflicts. Cleaver and de Koning, for example, point out the significance of other social contexts that are not necessarily rational or profit maximizing:

> »In securing access to resources and services, people are also concerned with wider ends (related to order and meaning, identity and citizenship, wellbeing). This means that practical arrangements for managing resources are imbued with wider social significance and can be traced back to the generative principles of the social field.« (Cleaver and de Koning 2015: 9)

The sociological approach of institutionalism considers the cultural development of institutions. Here, individuals are regarded as not fully informed in the sense of having an objective overview of their contexts, and, consequently, they are not able to make completely rational decisions. Thus, institutions step in which are formed according to cultural settings and make social acting and communication possible by providing structures. Therefore, they affect basic preferences and identity (Hall and Taylor 1996). Hall and Taylor underline that, according to sociological institutionalism, rationalism and win-maximizing strategies are still present in the actor's behavior:

»However, sociological institutionalists emphasize that what an individual will see as 'rational action' is itself socially constituted, and they conceptualize the goals toward which an actor is striving in much broader terms than others do.« (Hall and Taylor 1996: 16)

Peters (2005: 31) describes this socially influenced individual behavior as the »*logic of appropriateness*«, which means that, through institutions, individuals perceive guidelines about how to behave appropriately as corresponding to their position in society. Romzek clarifies the importance of informal institutions as he explains the *logic of appropriateness*:

»In this normative conception of institutions, the most important manifestations of the logic appropriateness are the mundane and routine standards of behavior. Hence, it eludes that not only are formal policies, laws and administration important for the existence of an institution, institutions may exist with non-formal appearances of norms. Norms are the result of shared notions of appropriate behavior in which institutions are to be founded upon. It is important that institutions share the same norms as society or else there would be a clash of norms.« (Romzek 2009: 30)

Important scholars of *New Institutionalism* are Douglass North and Mancur Olsons whose works are fundamental to the Common-Pool-Resource (CPR) theories. They are grounded on game theory and the assumptions of rational-choice. Although the scholars do not completely disregard the normative approach, they are still more related to the principles of rational-choice. For example, resource management is seen as decisions and actions that are made collectively in order to reach high efficiency and reduce risks. This decision-making process bears transaction costs. Thus, because of the utility maximizing nature of society as assumed by this approach, institutions are built in order to assure higher security of expectation and therefore, to reduce the costs. The researchers claim that political institutions set the »*rules of the game*« (North 1990: 3) and therefore, shape the behavior of the actors through »*collection of norms, rules, understandings and perhaps most importantly routines*« (March and Olsen 1989). Regarding the common property resources, »*Conflicts are seen primarily as distributive conflicts about the access to a certain amount of resource (e.g. water).*« (Houdret 2010: 68) The fundamental assumption that leads to this conclusion is the idea that scarce resources lead to collective action dilemmas. This means that individual decisions to optimize the individual's preferences. such as resource access, worsen the collective situation (Hall and Taylor 1996, p. 12). Hardin coined this theory in his publication »*The tragedy of the commons*« (Hardin 1968). Therefore, property rights have taken on a dominant role in the development of sustainable resource management (Hall and Taylor 1996: 12; North 1993).

Consequently, this school of thought has also influenced the question of water privatization. The idea behind that, which referred primarily to this theory, was to undermine this misleading individual behavior by handing the responsibility for the resource to the market. Because of the win-maximizing efforts, the most efficient handling of water would be reached only on the market, which will contribute to its protection (UNCED 1992, Briscoe 1996). Nevertheless, in recent literature scholars caution against the dangerous impacts on communities and their right to water (Boelens 2008; Boelens et al. 2011, Castro 2007; Getches 2005, Roth et al. 2005)

By analyzing a great amount of case studies, the political scientist and Nobel price holder Elinor Ostrom, developed an alternative institutionalist school of thought. She found that in contrast to Hardin's theory, groups of individuals are capable of successful common pool resource management and do neither depend on complete state control nor on privatization (Ostrom 1990). Prominent scientific approaches analyzing natural resources conflicts based on this strain are the eight design principles – mostly seen as a key to a good performance of governance on commons – and the Institutional Analysis and Development (IAD) framework (Ostrom 2005a), both developed by Elinor Ostrom and her colleagues. The eight design principles are characteristics that Ostrom found repeatedly in different case studies. Hence, they can be used as guidelines to successful common pool resource management. However, she stresses the importance of context in each case, so that these principles are indications rather than instructions. The eight principles are:

»1. Clearly defined boundaries
Individuals or households who have rights to withdraw resource units from the CPR must be dearly defined, as must the boundaries of the CPR ilself.
2. Congruence between appropriation and provision rules and local condidons
Appropriation rules restricting time. place, technology, and/or quantity of resource units are related to local conditions and to provision rules requiring labor, material and/or money.
3. Collective-choice arrangements
Most individuals affected by the operational rules can participate in modifying the operational rules.
4. Monitoring
Monitors, who activly audit CPR conditions and appropriator behavior, are accountable to the appropriators or are the appropriators.
5. Graduated sanctions
Appropriators who violate operational rules are likely to be assessed graduated sanctions (depending on the seriousness and context of the offense) by other appropriators, by officials accountable to these appropriators, or by both.
6. Conflict-resolution mechanisms

Appropriators and their officials have rapid access to low·cost local arenas to re-solve conflicts among appropriators or between appropriators and officials.

7. Minimal recognition of rights to organize

The rights of appropriators to devise their own institutions arc not challenged by external governmental authorities.

For CPRs that fire parts of larger systems:

8. Nested enterprises

Appropriation. provision. monitoring, enforcement. conflict resolution, and gov-ernance activities are organized in multiple layers of nested enterprises. « (Ostrom 1990:90)

Elinor Ostrom constantly worked on her approach to common-pool re-source management. In her later work she developed another framework called social-ecological-system framework (SESF) which focuses on the context-bound complexity in case study research and the interconnection of social and ecological factors (Ostrom 2009). The development of Ostrom's work from an institutionalist approach, based on game theory to a context bound socio-ecological framework, shows the connection be-tween both approaches as well as the flexibility of her work. Based on that debate, I chose to work with the SESF of Ostrom to embed the actors' be-havior in a broader context (cf. chapter 5.1).

Facing the before mentioned challenges concerning common resource governance, *Actor-centered Institutionalism (AI)* is an analytical frame-work that merges actor-centered and institution centered approaches. This approach is an agent-based model to explain collective behavior and was developed by Renate Mayntz and Fritz W. Scharpf in 1995.

»Agent-based models are appropriate for this task since they can include hetero-geneity of agent characteristics, agent-agent interactions, evolution of agents' characteristics, and learning. Agent-based models can also be used to explore pos-sible behaviour of not-yet-observed combinations of assumptions.« (Poteete et al. 2010: 192–193)

It follows the thoughts of *Sociological Institutionalism* as it defines social phenomena as a product of their institutional context (Scharpf 2006: 17). The AI can be used as a tool to analyze, evaluate and prescribe cases on the base of this frame (Scharpf 2006: 91). It qualifies conclusions about the processes of the development and emergence of political output. Still, it needs to be applied under the special conditions of each case. Therefore, it takes a kind of intermediate position between theoretical approaches and empirical case studies and tackles the criticism of high abstractness and hyped-up essence of agent-based models (Scharpf 2006: 81f; Poteete et al. 2010: 193).

A CRITICAL PERSPECTIVE ON INSTITUTIONALISM AND SOCIAL-ECOLOGICAL SYSTEMS

The human geographer Frances Cleaver (2012) labels the strains displayed above as »*mainstream institutionalism*«. In specific, she refers to the theories of »*common property and collective actions*« applied in *New Institutional Economics* (Hall et al. 2013) and the social-ecological system approaches and adaptive governance theories that were developed from those theories (Cleaver and Whaley 2018) Interestingly, she identifies an emerging school of thought and calls it »*critical institutionalism*«. This strain is not completely antagonistic to the common theories but amplifies them with a broader understanding of institutions – especially with regard to the application of institutionalism in field studies. Critical institutionalism points out the three main concerns: »*(a) the complexity of institutions entwined in everyday social life, (b) their historical formation; and (c) the interplay between the traditional and the modern, formal and informal arrangements.*« (Hall et al. 2013: 5) In other words, it

> »[…] offers the conceptual toolkit for illuminating process (how particular governance arrangements emerge and are enacted); power (how they are shaped to benefit some and not others); and meaning (how they become invested with meaning and so gain legitimacy and endurance.« (Cleaver and Whaley 2018)

The following part compares the insights of critical institutionalism to the institutionalist and socio-ecological system approaches shown above.

To a, complexity: Critical institutionalism criticizes the romanticizing of communities by questioning the independency of local institutions and by underlining interlinks of the institutions to a broader system (Hall et al 2013, Pigg 1996, Agrawal 2001). The scholars of this strain doubt »*the presumption that local practices are internally generated*« (Hall et al 2013: 8). Consequently, they claim that local institutions and the governance of local commons are influenced by national and international impacts (Komakech et al 2011; Hall et al 2013; Cleaver and de Koning 2015). This also means that scholars of critical institutionalism

> »question the underlying rational choice assumptions of much institutional thinking. Instead they emphasize the multi-scaler complexity of institutions entwined in everyday social life […]«(Whaley and Cleaver 2017: 57).

Likewise, Elinor Ostrom and her colleagues in her later works state that this conventional thinking – especially regarding the tragedy of the commons – is disproved by several case studies. They ascertain that sustainable use of common natural resources also occurs without the existence of private individual or of state property. And even if it does exist, it will not necessarily lead to sustainable use. In fact, tenure security and local monitoring systems run by the users themselves have shown to lead to more efficient collective decisions (Poteete et al. 2010: 45; Ostrom 2008: 13). As a consequence, scholars claim that the complexity and the context-dependency of water governance needs to be taken into account (Hall et al. 2013: 5–6; Pahl-Wostl et al. 2010: 572; Cleaver and de Koning 2015). Moreover, critical institutionalists claim that power and politics need to be considered in social and ecological system analysis. By that, they put a less optimistic view on the development of resilient structures (cf. Cleaver and Whaley 2018: 2). For example, while social-ecological system frameworks may regard powerful actors as change agents towards sustainable governance, the critical institutionalism lens focuses on the question of how those actors might take advantage of their power (cf. ibid.: 10). Moreover, is aims to discover not only individual distribution of power but also underlying power structures inherent to the institutional construct (ibid, cf. Whaley and Cleaver 2017: 59).

To b, historical formation: Furthermore, this strain criticizes ideologized politics based on Western concepts. As early as in 1993, North claimed that the *Rational Choice Approach* applies only to (some of the) »first world countries«. He preferred *Historical Institutionalism* to explain institutional change under the conditions of time and non-rational acting humans. The outstanding feature of the historical approach is its temporal dimension as it follows questions *»which seek to discover how institutions evolve through time and why institutions that produce poor economic (and political) performance can persist«* (North 1993: 3). Nevertheless, this approach of path dependency was blamed for being too one-dimensional and for focusing only on one section of institutional development (Walby 2007: 465). Regarding natural resources, critical institutionalists emphasize past trajectories. They claim that resource management is influenced by traditions and historical formed institutions that are independent from a chronological order (Cleaver and de Konig 2015: 6):

> »Departing from underlying rational choice assumptions of commons scholarship, critical institutionalists takes the view that resource governance systems are socially constructed, whereby meaning and social reality is historically and geographically situated and emerges from the interaction between members of a group or society (Berger and Luckmann 1967).« (Cleaver and Whaley 2018: 5)

Regarding this, Pahl-Wostl et al. see social learning as a key factor for institutional change in the water sector. They claim that *»[...] [it] has a strong influence on the nature of multiparty cooperation and social learning processes.«* (Pahl-Wostl et al. 2010: 573) However, the most common and prominent explanation for water scarcity is population growth. Other causes such as different levels of social adaptation capacities and resource dependency and their historical and cultural formation are mostly left aside (Houdret 2010: 72). Houdret claims:

> »The history of the relations between different parties, competition between sectors, villages or city quarters and immaterial expression of power constructions are essential for the development of conflict potentials as well as the negotiation of compromises and the avoidance of escalation - in the involvement of natural resources as well as other conflicts.« (Houdret 2010: 73)

The SESF by Ostrom recognizes historical formation as an indicator to explain the focal action situation (McGinnis and Ostrom 2014). Whereas the AI by Mayntz and Scharpf (1995) allows to draw deeper on path dependency because it considers action orientation developed from cognitive and motivational aspects such as identity and interest, and from orientation of interaction. Furthermore, this approach aims at a less abstract explanation and draws on empirical research. Hence, it opens methodological pathways to consider historical formation.

To c, interplay of formal and informal: The scholars of Sociological Institutionalism »*analyze the way institutions produce meaning for individuals, offering theoretical building-blocks for normative institutionalism in political science*« (Romzek 2009: 26). This approach assumes that in the practice the individuals' actions are guided by external influence (Schulze 1997). Individuals are influenced by the norms that they internalized due to their role in society. Moreover, institutions offer patterns of cognition and interpretation that give sense to the action of the individuals (Hall and Taylor 1996, Berger and Luckmann 1993). Critical Institutionalism takes this claim one step further because it criticizes common approaches for producing a too simplified outcome. According to Hall et al. (2013:6) »collective action scholars focus on efficiency and the health of the commons«. Therefore, they use institutionalism as »*a tool for researching community resource management and governance mechanisms and informing international donors and developing country policy-makers.*« (Hall et al 2013: 5). Contemporary scholars state that following this idea may have significant consequences for the communities as these interventions can threaten the informal institutions that are in some cases essential for a sustainable collective action concerning resources. Therefore, in a lot of societies research management and conflict solutions occurs on a traditional and informal level. Houdret and Shabafrouz (2006: 11), for example, warn against underestimating these informal factors, as they »might guide the long-term strategy of the actors involved, and determine their behavior to a higher degree than formal agreements.« Especially in water conflicts, formal institutions are often seen as a second option to the users. This diverging understanding of the law opens a gap between theory and practice (Houdret 2010: 73). Furthermore, Houdret and Shabafrouz (2006) underline that change in water regulation is a highly normative topic. The institutional context often reflects the existing power relations. Consequently, Houdret and Shabafrouz claim that the establishment of new institutions must consider the existing formal and informal rules and recognize their symbolic value in order to gain social and political legitimacy (Houdret and Shabafrouz 2006: 27). Also, Cleaver and de Koning state that:

> »Poor and marginalized people often find it difficult to shape the formal rules and the rules in use, to negotiate norms, and experience the costs and benefits of institutional functioning differently to more powerful people.« (Cleaver and de Koning 2015: 10)

Water privatization should serve as an example here. Generally, there are several requirements needed in order to introduce water trade and markets to a country. In addition to reliable and available data about usage rights, water quantity and systems of measuring the usage, new political and social rules must be established (Hoering 2006: 8; Perry et al. 1997: 12). At the same time, the existing values, norms and rules must be reconsidered. Houdret and Shabafrouz (2006) state that processes of water privatisation are mostly built on a formal level only. Hence, they are neither embedded into the informal context, nor do they possess social or political legitimacy (Houdret and Shabafrouz 2006: 11). Therefore, the »*institutional framework does not or no longer correspond to the actor's constellations and priority*«, and water management fails to be efficient (Houdret and Shabafrouz 2006: 26). As a response to that, this upcoming school of thought places »*emphasis more on the socio-economic equality and poverty reduction*« (Hall et al. 2013: 6). Cleaver as a main representative of Critical Institutionalism introduces *institutional bricolage*, a concept that seeks to overcome the distinctions of formal and informal, external and internal and that regards institutions as a mixture of socially or informal developed rules and bureaucratic, formal ones (Cleaver 2012, cf. Whaley and Cleaver 2017: 61):

> »The idea of institutional bricolage seeks to avoid the false distinction of portraying institutions as either clearly formal or informal, and emphasizes that local participants themselves, as well as intervening individuals and organizations, have some ability to shape institutions for managing resources such as water.« (Jones 2013)

Cleaver (2012: 45) defines institutional bricolage as:

> »The adaptive processes by which people imbue configurations of rules, traditions, norms, and relationships with meaning and authority. In so doing they modify old arrangements and invent new ones but innovations are always linked authoritatively to acceptable ways of doing things. These refurbishments are everyday responses to changing circumstances.«

Consequently, actors redefine constantly the institutions at hand while the institutions' resilience depends on the social fitting (cf. Cleaver and Whaley 2018: 6). Cleaver and Whaley emphasize that those newly formed institutions do not have a normative connotation:

> »Features related to institutional resilience may encompass adaptability, legitimacy, functionality, and endurance but these are assessed through the outcomes they produce, from a social justice perspective.« (Cleaver and Whaley 2018: 6)

Furthermore, Critical Institutionalism focuses on the concept of meanings that underlie the cognitive and motivational aspects of the individuals involved:

> »For critical institutionalists, it is impossible to understand environmental arrangements without appreciating that meaning and values adhere to them beyond their directly instrumental function. These meanings encompass worldviews about cause and effect in the human and natural worlds and different logics of action (for example the comparative values attributed to collective or individual action). Attribution of meaning is crucial for legitimizing and sanctioning relationships by relating them to accepted knowledge and familiar socio-political and environmental orders. Meaning (and power) therefore helps to ensure the acceptability and durability of institutional arrangements. Lessons for adaptive governance include the need to be aware that multiple processes of meaning making (beyond those of the adaptive governance focus) will likely shape adaptive governance arrangements in unplanned directions. » (Cleaver and Whaley 2018: 10)

Thus, scholars of critical institutionalism produce a more complex theory on commons that aims at providing a more realistic view on the study cases (cf. Cleaver and Whaley 2018). However, they seem to find fewer practical guidelines for policy-makers. For this reason, Hall et al. (2013: 32) call for the development and implementation of flexible concepts that give insights into the changing circumstances and dynamics of action on an individual and a collective level.

SUMMING UP

As shown above, common property research and institutionalism are developing research fields which contribute to sustainable resource management. However, the different schools of thoughts were discussed relatively isolated. As early as in 1996 Hall and Taylor stated:

»There is ample evidence that we can learn from all of these schools of thought and that each has something to learn from the others.« (Hall and Taylor 1996: 24). In this paper, I propose an approach that is ample enough to implement the mentioned contributions of critical institutionalism if necessary, but that provides guidelines and a structure which can be applied on an individual case study. More specifically, I take the SESF developed by Ostrom (2007, 2009) and her colleagues as a base to address complexity and historical formation. Then, this study deepens the analysis of the actors' behavior by applying the AI developed by Mayntz and Scharpf (1995) in order to address issues such as power and meaning. Recognizing the insights of institutional bricolage, this study uses the approach of social innovation studies to evaluate the possibilities of a broader institutional change. Both analytical approaches as well as their synthesis and connection to social innovation theory are presented in the next chapters.

5.1 Contributions of Ostrom's social-ecological system framework to this study

Regarding recent institutional development in the Chilean water sector, scholars state that further research on institutional changes and their connection to socio-ecological systems is needed (Clarvis and Allan 2014: 88). Retamal et al. (2013: 12) demand such academic efforts not only in order to create environmental and financial indicators for water issues, but also to generate stronger support by social sciences to analyse behaviour of the diverse actors and their motivations. Following those claims and in order to cope with the complexity of case studies, I propose the SESF developed by Elinor Ostrom (2007, 2009) and her colleagues.

Although scholars of critical institutionalism criticize common social-ecological system approaches for ignoring questions of social justice, political, cultural and sociological complexity, I argue that the SESF developed by Ostrom is broad enough to include those issues (cf. Cleaver and Whaley 2018). Ostrom and her colleagues have constantly developed this framework further. Thereby, they offer a diagnostic approach which can be adjusted according to the research question in terms of indicators and their weighting (Ostrom and Cox 2010). Nevertheless, the SESF developed by Ostrom is not the only approach that connects social and ecological systems. Binder et al. (2013) compared ten established frameworks and provide assistance in deciding which is the most suitable one for the respective research. Based on Binders et al. (2013), I will explain why I focus on the SESF. Binder et al propose three main questions:

> »[a] Do you study the effect of the social system on the ecological system, the effect of the ecological system on the social system, or are you interested in understanding the reciprocity of both systems?

> [b] How do you conceptualize the environmental system? Do you conceptualize it from the perspective of its utility for humans? Or do you want to understand it by itself?

> [c] Does the research question require an analysis or an action framework?« (Binder et al. 2013: 15)

To a) and b) In order to recognize the complexity of the case, the present study aims at understanding the interplay between both systems – the ecological and the social system. However, an anthropocentric approach is necessary because it combines the theoretical analysis with actor-centered institutionalism (AI, see next chapter). According to Binder et al (2013: 13), the SESF by Ostrom (2007, 2009) and the Management and Transition Framework (MTF) by Pahl-Wostl (2009; Pahl-Wostl and Kranz: 2010) are suitable in that case because both belong to the group of integrative frameworks. Binder et al. propose some research questions out of which one is very similar to the leading question of the present paper: *»What are the barriers and drivers for a transition toward sustainable water management in a catchment area?«* (Binder et al. 2013: 13–14). According to the authors, the SESF is the only one with which the social and the ecological systems can be analyzed on an almost equal level (Binder et al. 2013: 11).

To c) First and foremost, the present study is interested in understanding the structure of the social dynamics in the case of an environmental conflict. Secondly, the desired outcome is principles of how to improve the water governance. Consequently, an analysis framework fits the aim of the study. In that case again, Binder et al. (2013) claim that among others the SESF and the MTF are helpful. However,

> »Among these general frameworks, only SESF offers a generic data organizing structure. It is the most general framework, and the data collected within its structure could potentially be used in any of the other frameworks analyzed […].« (Binder et al. 2013: 12)

The SESF aims at establishing a common almost *»non-disciplinary«* language (McGinnis and Ostrom 2014: 1). This transferability allows for a higher range of comparability and has the advantage to provide the possibility of adding another analyzing method if necessary.

Ostrom and her colleagues used this approach on a variety of case studies concerning common goods such as water. They developed an overview of the potentially influencing features of resource and governance systems like actors and institutions (Poteete et al 2010: 236). They call these features the first, second and third (and so on) tier variables. The first tier divides the socio-ecological system into resource system and its units and governance system and its actors[10] and distinguishes between interactions and outcomes. The following tiers are more detailed variables that may influence an action situation (Poteete et al 2010: 235; Binder et al 2013; Ostrom 2007; McGinnis and Ostrom 2014: 1). However, all of these are (potentially) interconnected and influenced by the surrounding related ecosystems and social, economic and political settings.

While in the initial phase of the SESF the term »action situations« was represented by interactions and outcomes only, after the first criticism Ostrom decided to outline the importance of action situations more clearly and to include this term in the framework. This modification has led to a closer connection between the rather new SESF and the IAD[11] first introduced by Kiser and Ostrom in 1982. Both define action situation similar to the AI as a »(...) *situation in which actors in positions make choices among available option in light of information about the likely actions of other participants and the benefits and costs of potential outcomes.*« (McGinnis and Ostrom 2014: 4)

As mentioned before, the SESF is not completely fixed but a research tool which develops and improves through application and learning (McGinnis and Ostrom 2014; Basurto et al. 2013). Therefore, it has changed its categories and tiers over time. Basurto et al. (2013) claim that especially the implementation of tiers that reflect power structures and learning processes is understudied (Basurto et al. 2013: 1374). In 2014, McGinnis and Ostrom (2014: 9) offered an alternative list of second-tier properties for governance systems which are tackling the demand of

---

10   Initially, the Ostrom and her colleagues used the term »users«, but, as they found it to be too constricting in some cases, they changed it to actors in order to involve also individuals or groups that influence but not directly use a certain resource or governance system. Nevertheless, an actor can be classified as a user. (McGinnis and Ostrom 2014: 6)

11   »The IAD [Institutional Analysis and Development] framework is best thought of as a meta-theoretical, conceptual map that identifies an action situation, patterns of interactions and outcomes, and an evaluation of these outcomes«. (Poteete et al. 2010: 40)

Basurto et al. Most of these tiers, like *population* or *regime type,* are comparably easy to identify and help to structure research and to find indicators for specific research questions. Nevertheless, other features in this list, which are especially interesting for the present study, are more complex: In addition to other features, McGinnis and Ostrom added *Repertoire of norms and strategies* as a second-tier variable since they considered that »*norms per se have received less explicit attention by scholars working with either the IAD or SES frameworks.*« (McGinnis and Ostrom 2014: 10) Furthermore, they added *Network structure* which »*refers to the connections among the rule-making organizations and the population subject of these rules.*« (McGinnis and Ostrom 2014: 10) As the SESF is still developing, Basurto et al. suggest that a combination with another scientific tool would help to identify the missing tiers for an appropriate analysis. In this paper, SESF is combined with AI by Mayntz and Scharpf (1995). The following chapters display the AI and its syntheses with the SESF.

5.2 Actor-centered institutionalism in the light of water governance

The actor-centered institutionalism (AI) of Mayntz and Scharpf (1995) refers to an individual policy problem, to the actors involved, their constellations and the given modes of action and finally to the political output. According to the AI, institutional context does not only consist of aggregated interests of the individuals, formal rules, processes and norms but also recognizes the »*the symbol systems, cognitive scripts, and moral templates that provide the 'frames of meaning' guiding human action*« (Hall and Taylor 1996: 14). Consequently, institutions rely on a common acceptance within society. The AI recognizes that by facing the common criticism about disregarding the interplay between the formal and the informal. The AI goes even deeper and takes in a multi-stage perspective where the institutional framework influences the actions of organizations which then shape the institutional framework for the individual actors' behavior (Mayntz and Scharpf 1995: 44). Thus, AI works with a tight definition of the term institution. It distinguishes between actors and institutions while the latter are dependent or independent variables which enable communication and collective action but do not necessarily determine it (ibid.). Concerning water governance, formal institutions are for example official rules and laws about water regulation. This will be illustrated by some examples. Saleth and Dinar (2005) as well as Gopalakrishnan (2005) divide formal institutions into three sectors: water law, water policy and water administration. The performance of each can be measured by its physical outcome (e.g. water quality), financial outcome (e.g. water prizes) and equity outcome (e.g. accessibility to water). Informal institutions are established through customs and habits and exist in the context of water governance in for example negotiations and arrangements (Houdret and Shabafrouz 2006).

Furthermore, AI takes the broader context of each case into account. It analyzes the institutional context interlinked to its external environment (Mayntz and Scharpf 1995). By doing so, AI approaches the complexity of each case – a point of interest that is missing in the classical institutionalism methods. According to the AI, the external environment »*encompasses all the issues not directly related to the water sector but which nevertheless represent constraints for the scope of action for water governance.*« (Houdret and Shabafrouz 2006: 12) Examples concerning water governance are the economic globalisation, resource scarcity caused by climate change coupled with a rising demand, or severe financial difficulties of the public sector (Houdret and Shabafrouz 2006). In the following subchapters, I will explain the different features of the approach by using examples of the water sector.

5.2.1 Actors

Since agent-based methods analyse behaviour mostly on a higher level than on the micro level, it is important to identify and categorize actors that follow the same aim such as access to water (Scharpf 2006). AI divides actors into three different types (Mayntz and Scharpf 1995: 49–51):
- Cooperative actors: capable of acting in formally organized majorities such as organizations, typically top-down structures (Scharpf 2006)
- Collective actors: a majority with the same action orientation of its members but no formal organization, for example coalitions, social movements, clubs and associations. Scharpf (2006) further divides this group by their control of their action resources and their goals.[12]

12　Here, generally coalitions have separate goals and control systems, social movements have collective goals but separate control of action resources, actors in a club may have separate goals but collective control of action resources and associations have collective structure on both (Scharpf 2006: 102).

- Other social elements or groups: persons that is unified by a certain action relevant feature such as families, or networks. They are called aggregated actors.

However, scholars that work in the field of water governance have found different ways of classifying actors. Houdret and Shabafrouz divide the actors in two ways. On the one hand, by their scope of geographic acting from local to international and, on the other hand, by their legislative body – be it political, civic or economic actors. On an international level, there are different financial and legislative institutions involved in the water sector – such as the World Bank or the International Monetary Fund, or the environmental standards set by the OECD (Houdret and Shabafrouz 2006: 17). Houdret and Shabafrouz divide the actors in governmental authorities, water users and water providers. Nevertheless, within these different actor groups, there are again distinctions to be made. To illustrate the classification of actors of the AI, the following Table 1 shows a simplified model of the actors in water governance. The illustration is based on Houdret and Shabafrouz (2006) and combined with the distinctions from AI using the example of the case study's actors in Petorca (Chile). The categories of AI are shaded in grey.

Table 1: Types of actors in water governance. Categories of AI shaded in grey. (Derieved from Hourdret and Shabafrouz 2006, Scharpf 2006 and modified by Roose)

Actors	Political actors	Civil society	Economic actors
International	Policy Networts	International organizations and foundations,	Bank and funds *Cooperative actors*
National/ regional	Governmetal authorities *Cooperative actors*	NGOs, Movements & Activists *Collective actors*	Water provider *Cooperative actor*
			Industrial users (e.g. big agricultural and mining companies) *Cooperative & collective actors*
Local	Municipalities, local authorities *Cooperative actors*	Water users *aggregated actors* Water User Organizations (WUOs) *Collective actors*	(e.g. small-scale farmers) *Aggregated actors*

According to the AI, all these actor groups are formed and influenced by the institutional framework. Nevertheless, it is important to mention that the institutions do not determine the actors' behavior on an absolute level. Firstly, norms, rules and values can be defined and, secondly, institutions can determine the availability of natural and technical resources only to a certain limit (Mayntz and Scharpf 1995: 49).

5.2.2 Action orientation

According to the AI, Action orientations are motivations and interests of actors that lead them to their individual usage of the given scope of action. By that, this approach dissociates from the rational choice theories. Scharpf (2006: 110-111) states that the actions of individuals are led by certain mind-sets. These are filled with (subjective) preferences which are not consistent in every situation or between different temporal phases. Pahl-Wostl et al. (2010) call this the *»Mental Models«* of actors that determine the criteria which with actors analyse and perceive the water system. Observing and collecting these data is an elaborate task. For this reason, the AI clusters the action orientations into simpler parts which can be explained more easily due to institutional settings or empirical research (Scharpf 2006: 111). In order to describe the action orientation, it is especially important to identify the social unit of the actor since this unit may determine the actor's perspective. However, even in the case of cooperative actors the action is done by individuals. Therefore, it is important to empirically ascertain the social units involved in each case (Mayntz and Scharpf 1995: 52–53). Moreover, Mayntz and Scharpf identify three leading influencing factors for the action orientation:

- Cognitive aspects: the perception of the situation and its causality, in other words, the situation's interpretation
- Motivational aspects: driving factors or criteria for certain decisions, distinguished between: interests, norms and identity
- Orientations of interaction: Mayntz and Scharpf claim that this factor is especially important for the analysis as it gives the opportunity to classify relationships of actors into certain types: hostile, competitive, selfish-rational or cooperative (Mayntz and Scharpf 1995: 57)

These three factors reflect the impacts on human behaviour identified by the critical institutionalism. Action orientation is influenced by both, the dependence on a social system (compare to social unit) and the creativity of the actor's identity (compare to motivational aspects). Additionally, action orientations are shaped by power dynamics and are embedded in routinized practices. Therefore, the actor may not be completely aware of factors such as cognitive aspects (Cleaver and de Koning 2015: 8).

5.2.3 Action situation and scope of action

As already mentioned, the behaviour of an actor depends on many different factors. Furthermore, actions are based on the given situation (Mayntz and Scharpf 1995: 59). According to the AI, a situation is determined by the »*action relevant, social or not social conditions of the environment of a particular actor*« (Mayntz and Scharpf 1995: 58). The pertinence of an action depends on the situation's incentive character (such as threatened loss or possible profit) and the options to act possible in the given situation. The latter can be either institutional or not. The access to resources, for example, may be restricted by institutions such as laws or by geographical conditions. The perception of the environment and other factors of the action situation then define the success of the strategy chosen by the actor (ib.). In the water sector for example, Houdret and Shabafrouz (2006: 20) claim that options for the national governance to act are limited due to the great influence of international financial institutions. The leeway left to the actors is called scope of action. In the case mentioned by Houdret and Shabafrouz, the scope of action left for the user groups relies on the legitimacy of their actions and on their technical or legal knowledge. Moreover, Pahl-Wostl et al. (2010) stress that in every situation, the actors take on certain roles. They use knowledge which depends on the situation and is therefore called »*situated knowledge*«. This connects the actors to the action orientations and the action situation:

> »A 'role' has as attribute a range of possible actions and entitles actors holding this role to certain knowledge. It is assumed that actors activate 'Situated Knowledge' within the context of a specific 'action situation.' It is, therefore, linked to an actor and the situation. Situated knowledge captures the importance of framing and reframing and the embeddedness of knowledge in a social context. The 'Observed State of Water System' is a specification of knowledge used in an 'action situation' to evaluate the state of the 'water system'. The choice of which factors and criteria are what is used reflects the perception of 'actors' about what is important for them to make a judgment about their individual satisfaction and the achievement of 'management goals'.« (Pahl-Wostl et al. 2010: 576)

5.2.4 Actors constellations and modes of action

The behavior of actors in a specific situation is the overall result of complex action interdependencies. Beneath preferences, perceptions and social units of every individual, actions also depend on the present actors around because one actor alone will most probably not be able to determine any political outcome with his or her own resources (Scharpf 2006: 87). To structure this interrelation, the AI outlines two factors: the actor constellations and the ways of social action coordination, called modes of action. Depending on the institutional context, a higher or lower amount of opportunities of how to make decisions is given. Mayntz and Scharpf classify the modes of action into: unilateral action, negotiated agreement, majority vote and hierarchical direction (Mayntz and Scharpf 1995: 60). Those modes of action are influenced by the existence or non-existence of institutions in terms of rules, values etc. In each context, actor constellation and modes of action may differ independently. For example, while an anarchic field only allows unilateral action and, in some circumstances, negotiated agreements, the scope of action in hierarchal organizations like the state can be much broader – including majority vote and top down processes (Scharpf 2006: 91).

This should be explained by an example. In the anarchic field few to no institutional rules exist. Action incentives are based on the aim of individual win-maximization. Trust between actors is typically low and interaction orientation is rather selfish-rational, competitive or hostile than cooperative. Because of that, a collective decision making is unlikely to happen (but not impossible, cf. Ostrom 1990). Consequently, options such as majority vote or hierarchical direction are provisionally cancelled. Taking the example of rule compliance (e.g. regarding water regulation) each actor needs to agree, but, if hierarchical direction could be implied e.g. through enhanced monitoring and sanctioning or constitutional change, the institutional framework would change and allow enhanced modes of action. In the light of Water Code reform debates this means, that the institutional framework should not be considered as fixed, but as modifiable.

Scharpf (2006: 92f) asserts that the combination of the information about interaction modes and actor constellation will lead to the determination of the problem-solving ability of the institutional system. For example, Houdret and Shabafrouz (2006: 25) state that globally current debates about modes of action in water governance are highly polarized between proponents of water privatization (following the hierarchical direction) and proponents of collective decision making (as in decentralised public management systems). Therefore, they do not represent the actual and more complex institutional context. Situations like that *»[...] contribute to a large degree to the current blockades on water governance.«* (Houdret and Shabafrouz 2006: 26). Also, Poteete et al. (2010) confirm these arguments as they note:

»Policy interventions that threaten informal rights often generate considerable opposition. Such clashes may be more frequent with multiple-use resource systems, where governments tend to prioritize commercial uses while overlooking or discouraging subsistence use.« (Poteete et al. 2010: 49)

Accordingly, the following question needs to be answered: how to find out about actor's constellations and modes of action? Scharpf (2006: 93) proposes three steps. First, the identified policy problem regarding the actor constellations and the actors that are indeed involved into the political processes must be scrutinized. This will lead to a very abstract but precise understanding of how the actors and their preferences will react to possible outcomes. Therefore, it will identify the conflict level, but not the difficulties of problem-solving strategies. For that, it is necessary to look at the modes of action. Thus, in the case of a certain actor constellation the political outcome can be varied and hopefully improved if the modes of interaction change and vice versa. The following subchapter shows how I have merged the AI with the SESF in order to create an analytical framework suitable for the present research case. Furthermore, it displays in how far insights of game-theory are used to explain the results of that approach.

5.3 Syntheses and the role of game theory and social innovation

To tackle the research questions of how the institutional framework impedes or facilitates sustainable water governance in the context of a water conflict, the SESF works as a guidance towards the potentially important variables and is combined with the AI. The AI then enables a more in-depth analysis of these features. This combination, I assume, will respond to the calls of critical institutionalism as it will cope with the historical formation, complexity and interplay of formal and informal institutions.

The AI analyses how actors behave in a certain collective action situation and why or why not a conflict or action situation permits certain forms of interaction. In order to do so, this approach needs information about the actors, their orientations, their scope of action etc. as explained before. The tiers of the SESF help to gather all the information necessary to describe the water conflict and the impacts of its surrounding features, which are the resource system, its units, the social, economic and political setting, the related ecosystem, the governance system and the actors. This information describes the »*focal action situation*« which in this study is the water conflict in Petorca. The information is then combined with AI in order to explain its coherence and how it influences the actors' behavior and the possible solutions of the described conflict.

CLARIFICATIONS AND ADDITIONAL INDICATORS

It should be mentioned that AI uses the term actors when referring to rule-making organizations like NGO's, private-sector organizations, while in the SESF those are classified as part of *Governance System*. In the present study, I took the categories of the SESF as a first step and then set them into the context of the AI. This means, I did not separate the actors into the two categories »Governance System« and »Actors« but rather took the different approaches as a chance to first categorize them with regard to their organizational structure according to the SESF and then classify them into »cooperative, collective and others« and analyze their behavior as proposed by the AI.

As explained in chapter 5, path dependency can play a significant role in explaining collective behavior towards commons. The SESF establishes a structure for coping with time and path dependency by introducing the second-tier variable »*historical continuity*« that tier

> »(...) is included to distinguish between systems of governance that have been in place for long periods of time and those that are more recent in form. All forms of governance have deep roots in historical precedents, but some systems are more inclined toward stasis and others toward more flexible modes of response.«
> (McGinnis and Ostrom 2014: 10)

Hence, this tier allows to incorporate certain learning processes in terms of water governance development (cf. Basurto et al. 2013: 1347). In this study, it indicates basic water governance development from 1951 to 2016.

In addition, insights of social innovation approaches were used to analyze real-political outcomes of the focal action situation. Those approaches focus on real-political case-study experience and provide useful indicators to evaluate institutional changes on a local level. Therefore, I argue that the concept of social innovations offers a more applicable tool for recognizing those circumstances which are called institutional bricolage on a community level by critical institutionalism. In other words, it enhances the list of indicators that can be added to the final discussion. Based on the research in social innovation by González et al. (2010: 57–64) and Pradel Miquel et al. (2013: 159-160) the following five indicators were detected and included into the discussion:

- Aims and visions
- Organizational structure
- Development
- Problems
- Impact on water governance

THE ROLE OF GAME THEORY

The chapters above show that »*The actor-centered institutionalism runs the risk [...] to become over-complex and to force a historical reconstruction due to the integration of institutional and action theory perspectives.*« (Mayntz and Scharpf 1995: 66) Therefore, Mayntz and Scharpf (1995: 66) propose to integrate the »*rule of decreasing abstraction*« by Lindenberg (1991). With regard to AI, this means to explain phenomena referring to actors only if they cannot be explained by institutions. Additionally, explanations for the actors' behavior should only be proved empirically if the observed actions cannot be explained by assumptions. Consequently, Mayntz and Scharpf (1995: 62) state that the AI should be seen more as a theoretical framework than a research method and recommend implementing game-theory to analyze interactions and network analysis to explain complex actor constellations. Game theory can be used to prove theoretical assumptions in experiments and has been applied in conventional common resource investigation. Also, a great part of Ostrom's work explaining behavior of actors in common pool resource management relies on the insights of game-theory. According to her, game-theory shows that trust and reciprocity are the building blocks of cooperative behavior (Ostrom and Walker 2003, cf. Messner 2012).

Hence, to answer the research question of how an institutional framework impedes or facilitates sustainable water governance, the question of Ostrom's work: » [...] how do groups of individuals gain trust?« (Ostrom 2003: 19) is of central importance to this study. Beckenkamp explains the connection of trust and institutional construct by highlighting the appropriateness and handiness of an institutional system. According to him, an institutional construct does not only need to provide structures that allow cooperation, but also its functionality needs to be applicable and understandable to the actors involved. He states that trust is psychologically as well as institutionally dependent and, therefore, an appropriate and good institutional design enables humans to have mutual trust while an inappropriate institutional design may severely disturb interpersonal trust (Beckenkamp 2014: 54).

Ostrom connects physical, cultural and institutional variables to trust and claims that trust influences reputation and reciprocity which then raises the cooperation level. A higher cooperation level will then lead to greater network benefits. More specifically, Ostrom detects four basic mechanisms that strengthen trust between actors in common-pool resource management and therefore potentially facilitate cooperative behavior:

- (a) Communication as direct as possible
- (b) Options to sanction opportunistic behavior
- (c) Inner, experienced heuristics, norms and rules (facilitate or impede cooperation)
- (d) Orientation towards reciprocity (positive behavior generates positive behavior and negative behavior generates negative behavior) (Ostrom 2003; cf. Messner 2012)

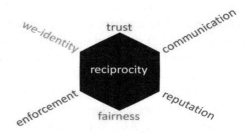

Figure 3: The cooperation hexagon. (Adapted from Messner et al. 2013: 15)

Messner et al. (2013) enhanced these mechanisms, respectively distinguished them in greater detail. They found seven mechanisms and visualized them in a hexagon with reciprocity being in the center (see Figure 3). The mechanisms defined by Ostrom (a) communication and (d) reciprocity are represented in that hexagon. The mechanism (c) inner, experienced heuristics, norms and rules are divided into »trust«, »reputation«, »fairness« and »we-identity«. Furthermore, Messner et al. (2013: 15) named one mechanism »enforcement« which according to them means that correct behavior will be encouraged by positive (»praise« for »good« behavior) or negative (»punishment« for »bad« behavior) reinforcements. In that sense, this model is an advancement of the mechanism (b) options to sanction opportunistic behavior developed by Ostrom (2003; cf. Messner et al. 2016: 52).

However, game theory is highly criticized for being too simplistic and thus having little external validity (Potetee et al. 2010: 169). The assumptions made in these experiments are based on the theory of »the tragedy of the commons« and assume a win-maximizing nature of each actor the disadvantages of which for common resource investigations have already been explained above (cf. Houdret 2010: 68). Moreover, Hall and Taylor noted:

> »The usefulness of the approach is also limited by the degree to which it specifies the preferences or goals of the actors exogenously to the analysis, especially in empirical cases where these underlying preferences are multifaceted, ambiguous or difficult to specify ex ante. [...] The drawback, of course, is that this advance comes at the cost of conceptualizing intentionality in terms of a relatively thin theory of human rationality.« (Hall and Taylor 1996: 17–18)

Hence, the transferability of the outcome to real cases is limited as in reality systems are highly sensitive to even small changes (Walby 2007: 455; Potetee et al 2010: 169; Hall and Taylor 1996: 17-18). To overcome this abstractness and »artificial nature« of such agent-based models, Poteete et al (2010: 193) recommended combining this approach with case study research. Such an integrative approach was also developed for this study. It combines a case-study research with the SESF by Ostrom and the AI. The present research case will not be transferred into an abstract game as it would oversimplify the focal action situation; however, insights of game-theory, in specific the key trust building mechanisms, are used to explain and to discuss the identified results (see Part 4).

SUMMARY

The following Figure 4 shows, using a visualization of the Ostrom SESF, how AI is inserted into this frame, so that both complement each other. The outer frame represents the SESF by Ostrom, including the boxes in the corners representing resource systems and units on the left side and governance system and actors on the right side. All of them influence the focal action situation in the middle. Inside that frame there is a second box that represents the AI. It concentrates on the interplay of the institutional environment and helps analyzing the focal action situation by defining actors' orientation, situation and modes of action (see Figure 4).

Socio-ecological system framework (Ostrom 2007, 2009)

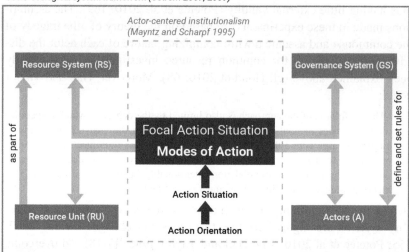

Figure 4: Synthesis of SESF and AI. (Adapted from McGinnis and Ostrom 2014: 4 modified by Roose based on Scharpf 2006)

The conceptualization was based on the framework-theory-model definitions made by Ostrom and her colleagues. According to them, a framework identifies many relevant variables and potential relationships, while a theory sets just a few assumptions and predicts outcomes; a model specifies the consequences and provides the opportunity to examine them (McGinnis and Ostrom 2014: 1–2). For the present study, the analytical framework is the synthesis of SESF and AI. The theory applied is based on sociological institutionalism and Critical Institutionalism (cf. chapter 5). Both analytical frameworks SESF and AI, rely on respectively recommend game-theory to explain the underlying mechanisms of the detected focal action situation. Consequently, insights of game theory were used to explain the outcome of the framework application.

6. Hypotheses

Water crises such as water scarcity are recognized as one of the main problems for humanity in the future (WHO and UNICEF 2017; World Economic Forum 2018). The triggers for this threat are ecological as well as sociological. Chile gives a vivid example for both factors. Despite its huge water availability, Chile is already suffering from water scarcity today due to its geographical conditions. While the south benefits from high precipitation and river flow, central and northern Chile have repeatedly struggled with drought. In the future, Chile is said to be one of the five countries worldwide suffering most from a drastic increase of water stress (Maddocks et al. 2015). However, it is widely recognized that water crises are rooted in and need to be dealt with (water) governance (UNESCO 2006). Chilean water governance is known as an outstanding example, mostly because of its water privatization. While some scholars argue that this system has led to a satisfactory coverage of water provision and contributed to economic growth (cf. Molinos-Senante 2018; UNESCO 2017; Saleth and Dinar 2005; Valdés-Pineda 2014; Naciones Unidas, Gobierno de Chile 2010), others criticize the system for being environmentally unsustainable and socially unjust (Bauer 2015; Retamal et al 2013, Belmar et al. 2010; OECD 2015; OECD/ECLAC 2016) . However, there seems to be common ground on the conclusion that water governance needs to be more inclusive in Chile concerning citizens involvement. This claim of inclusive water governance is widely recognized in science and in global policy. However, international agreements do not seem to reach local ground or do so in an unsatisfactory way. For example, although Chile consented to assure proper sanitation and water provision at several international agreements, in the central-region citizens suffer from the lack of water quantity and quality. In the Petorca province, citizens compete with a large-scale agricultural and mining industry over water because there is no prioritization of water for human consumption (INDH 2012). This phenomenon of divergent interests leads to water conflicts all over Chile. At the same time, social movements and other new creative strategies have emerged that aim at sustainable solutions to the problem. This study raises the question which institutional system facilitates or impedes sustainable water governance and if new creative strategies could pave the way towards sustainable water governance.

Based on current research, this study assumes that sustainable water governance must integrate actors in order to explore the benefits of participation. In this study, sustainable water governance is defined as an integrated management system that (a) recognizes the interplay of political, social, economic and administrative systems concerning water that (b) considers correlating systems outside of the water sector and (c) aims for adaptive strategies towards long-term environmental and social resilience by (d) applying strategies that increase the benefits of participation and cooperation.

Institutionalism aims at explaining human behavior by formal and informal institutions such as rules and values. This study uses an institutional perspective and takes Chile, respectively Petorca, as a research case. More specifically, it uses indicators of SESF by Ostrom (2009; McGinnis, Ostrom 2014) and AI by Mayntz and Scharpf (1995) as an analytical framework and states two hypothesis which have been developed by combining the theoretical framework and the given information about the research case. Following that school of thought, integration of actors, as demanded by sustainable water governance, presupposes cooperation which is influenced by the institutional framework (formal and informal; see chapter 5.2.4). Given the attested low state power in Chilean water governance, unequal water access and water privatization, the first hypothesis is:

> *Market-based principles dominate the institutional framework of water governance in Petorca and restrict cooperation towards sustainable water governance. (Hypothesis 1)*

Recognizing new creative strategies that have emerged in the region and the importance of local, community-based solutions in water governance policy, the second hypothesis argues as follows:

> *Social innovations overcome the market-based institutional framework and enhance cooperation towards sustainable water governance. (Hypothesis 2)*

The following part demonstrates how these hypotheses were generated in more detail by projecting the political problem of water scarcity in Petorca on the actor constellations in order to outline possible modes of interaction (Scharpf 2006: 123).

In the case of Petorca, the political problem concerns the scarce water resource and its management with regard to property rights systems, extraction and distribution. According to Elinor Ostrom, water is a common-pool resource due to its characteristics: a good that everyone needs, rival and hardly exclusive or unexclusive (Ostrom and Schöller 1999: 38). Since water is a scarce resource, mismanagement could lead to overuse or even destruction of it (tragedy of the commons, Hardin 1968). Thus, the given problem of this paper can be defined as a collective-good problem in which:

> «individual action produces negative or positive effects for others that will be disregarded by purely self-interested actors - which means that purely unilateral action would produce more negative and fewer positive effects than would be welfare optimal. If this process went unchecked, the result could be a Tragedy of the Commons (Hardin 1968), in which common resources are exploited and ultimately destroyed by rational self-interested actors. (Scharpf 1997: 90)«

Looking more closely at the underlying conflict, this study argues that the collective-good problem consists of an appropriation problem with existing insecurities about the allocation of spatial and temporal access, as well as a provision problem, in which conflicts about maintenance rules can be observed (conflicts about extraction rate or technical solutions for water withdrawal) (cf. Ostrom 1990: 46ff). However, problems of collective goods are strongly connected to problems of distributions, and their solutions are normative. In that sense, it needs to be defined normatively *»what should be considered a problem and what would constitute a good solution.«* (Scharpf 1997: 13) Consequently, it is essential to find strategies that scientifically deal with the problem of normativity. Scharpf suggests tackling this issue by recognizing its dependence on two dimensions: welfare production and distribution (Scharpf 1997: 15). Generally, there are two predominant welfare criteria: Kaldor *»[...] according to which policy choices are minimally acceptable if the gains to the winners are high enough to permit full compensation to all losers [...]«* (Scharpf 1997: 91) and Pareto

> »[...] which favors outcomes that improve value production in some respect (favoring particular types of values, at particular times and places, and benefiting particular persons) without reducing the satisfaction of any other value aspect.« (Ibid. 90)

Both concepts ignore the question of distributive justice and the plurality of its criteria (e.g. equity, equality and need amongst others, cf. Deutsch 1975; 1985). Scharpf states that it is not the researcher's role to judge which individual policy choices are right. However, the researcher should focus on »[...] the capacity of policy systems to reach good choices« (Scharpf 1997: 91):

> »The implication is that certain types of policy systems are generally capable of dealing with specific types of problems - and generally incapable of dealing with certain other types of problems - in ways that could satisfy the dual standards of welfare production and distributive justice.« (Scharpf 1997: 15)

Thus, it needs to be analyzed which welfare criteria would lead to which outcome concerning the problem of the collective good water. Underlying the Pareto criterion to welfare means that no actor or actor group should experience disadvantages through policy changes. Prima facie, it seems to be a more reasonable criterion for welfare production in problems of common goods than Kaldor. But, despite the potential gain, Pareto also faces normative problems of distributive justice, as

»[...] once the opportunities for such »costless« gains are exhausted, modern welfare economics declares itself incompetent to judge the distributive issues involved in trade-offs among different value aspects on the »Pareto frontier«. (Scharpf 1997: 90).

Moreover, the starting situation of the different actors may differ considerably. In the case of Petorca, small scale farmers and citizens suffer from low water access. Increasing their water access would probably affect large-scale agricultural companies negatively and vice-versa. Hence, the Pareto optimal outcome has already been reached and cannot solve problems of distributive injustice. In that sense, it is an unsuitable criterion if aiming for sustainable water governance.

Underlying the Kaldor criterium, the focus is on the overall welfare production so that losses of one actor need to be compensated by the wins of others. One could argue that this premise has already been aimed at in Petorca; that the economic win, for example, of large-scale agricultural companies is supposed to compensate for the loss of small-scale farmers and others by providing jobs and tax income. Therefore, big entrepreneurs are supported by politics (subsidies of water installation etc.). However, recognizing water as a human right implies that it can only be compensated with other values if the basic need is covered. Considering the distributive injustice, the social consequences and the dissatisfaction of the population, I argue that the intentions to compensate those losses were not successful because water management policy aimed at the Pareto criterium. Hence, Kaldor remains as the welfare criteria to be aimed for. Therefore, the following questions need to be addressed: Is a shift towards a Kaldor welfare production possible under the current conditions? Which is the current scope of action and which institutional restructuring would a change imply?

HYPOTHESIS 1:
According to Mayntz and Scharpf (1995), the institutional framework determines the variety of action modes that might lead to a sustainable water governance. Hence, to describe possible modes of action, the current governance style needs to be identified. Based on the current research that attests low governmental power in terms of legal regulation as well as compliance and based on the water privatization that aims to regulate appropriate resource use, I argue that water governance structure in Petorca is characterized by minimal institutions (cf. Scharpf 1997) following neoliberal principles of a free market (cf. market style governance by Pahl-Wostl (2015) as described by Meuleman (2008)). These rules of competi-

tiveness and cost-driven motives have expanded from the economic market and have become guiding principles inherent to the institutions involved. In line with this institutional framework and following the AI approach, potential modes of action are limited to one-sided or mutual adjustment and negotiated agreements (Scharpf 1997: 147, cf. chapter 5.2). However, to conduct »solution-capable« negotiations, an institutional construct is needed that allows additional modes of action in comparison to the current classical governance style of market mode. Under the conditions described, the actors are not capable of reaching a higher common welfare goal by cooperation. Based on that, the first hypothesis in this study is:

> *In Petorca market-based principles are predominant in the institutional framework of water governance. In this context, cooperation between actors is low and possible modes of action towards sustainable water governance are strongly limited. (Hypothesis 1)*

Testing hypothesis 1 aims at displaying the underlying mechanisms (M_1) between institutional framework and cooperation respectively trust level.

HYPOTHESIS 2

Consequently, a change towards a more complex institutional construct that enhances trust and cooperation such as a network system seems reasonable (cf. Pahl-Wostl 2015). To answer the research question »How does the institutional framework influence sustainable water governance?«, it is crucial to explore how actors develop trust (Ostrom 2003).

In Chile (inter-)national pressure on authorities to shape water management in a more inclusive and participative way is increasing (cf. OECD 2015, OECD/ECLAC 2016; chapter 4). In situations like that, Pahl-Wostl warns: »*Authorities often include participation to comply with legal prescriptions but without deliberate consideration of the potential benefits of a network governance approach*« (Pahl-Wostl 2015: 93). The statements of various Chilean scholars approve Pahl-Wostls assertion. Thus, participation is claimed to be seen as a challenge that needs to be tackled rather than a tool for development (cf. Retamal et al. 2013, Belmar et al. 2010; Espinoza 2013). The Chilean sociologist Vincente Espinoza frames it like that:

> »Public sector participatory initiatives, no matter how well intended, seldom go beyond the target population of specific policy actions. Even when public institutions create consultation committees involving users of the services, they tend to implement exhaustive screenings to avoid »disruptive« practices.« (Espinoza 2013: 405–406)

On the way to a more sustainable and inclusive water governance, it is important to recognize that an abrupt, radical change of rules would imply high transformation costs. For example, a (re-)nationalization of water rights would imply expropriations which are bound to high bureaucratic burdens and other social and economic consequences. Ostrom recommends prioritizing institutional change in an incremental process (Ostrom 1990: 141).

> »What is presumed to be a second-order dilemma, in which institutional change is viewed as one large step, may or may not have the structure of a dilemma when institutional change is viewed as a sequential and incremental process. The net payoffs of solving a small part of a large second- or third-order problem may be sufficiently high and distributed in such a manner that some participants will voluntarily provide initial second-order collective benefits, whereas they are unwilling to provide first-order solutions on their own. Solving some initial second-and third-order problems can help participants move toward solving first-order problems, as well as the more difficult second- and third-order problems.« (Ostrom 1990: 141)

Furthermore, she argues that actors experiencing successful implementation of small institutional change will learn stepwise and by this, successively gain social capital. Then, this expertise can be used »*[...] to solve larger problems with larger and more complex institutional arrangements.*« (Ostrom 1990: 190).

In Petorca, efforts towards a more participative water management exists besides the traditional forms of organization (cf. chapter 8.2). New creative strategies growing from within the local network are potential drivers for trust and reciprocity (building blocks of cooperation (Ostrom 2003, cf. Messner 2012, Messner et al. 2016). They can lead to a change of rules and incrementally change the institutional construct. Based on the benefits of cooperation and incremental change described above, this study argues that new creative strategies growing from within the local network are potential drivers for trust and reciprocity that could lead to a change of rules and, hence, an incremental change of the institutional construct. In this study, these strategies are called social innovations.

Figure 5 displays the assumption of social innovation enhancing trust and reciprocity, which then enhances cooperation. Cooperation facilitates the development of further new institutions or new combinations of existing ones (cf. institutional bricolage, Cleaver 2012) that enhance trust and reciprocity. Hence, social innovation may initiate an upward spiral towards incremental institutional change (cf. Figure 5).

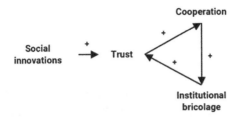

Figure 5: Upward spiral of social innovations. (Own illustration)

Applying this assumption to the present research case of Petorca, leads to the second hypothesis:

Social innovation overcomes the market-based institutional framework and enhances cooperation towards sustainable water governance. (Hypothesis 2)

By verifying this hypothesis, it is crucial to establish in how far the institutional framework has allowed these innovations to rise and in how far the institutional framework promotes or hinders them. In other words, the mechanisms of social innovation (M_2) that overcome the cooperation limiting mechanisms of market-based institutional framework (M_1) must be filtered. Moreover, it is crucial to discuss the range of actors involved and the social innovations potential radiance to enhance sustainable water governance.

Part 3: Results

Against the background of globally increasing water conflicts and crises, this study raises the question of which institutional framework facilitates or impedes change towards sustainable water governance (cf. Part 1 and 2). To answer this question, I chose Petorca, a region of the Chilean central-north, as a research case and developed an analytical framework based on the social-ecological system framework (SESF) of Ostrom (2007, 2009) and actor-centered institutionalism (AI) by Mayntz and Scharpf (1995). Based on current research and information about Petorca, I formed two hypotheses. The following chapters of Part 3 show the outcome of the application of this analytical framework to the research area. It starts by demonstrating how the analytical framework was applied and then shows the results going from a broader picture (social, economic and political setting; related ecosystems; resource system; governance system) towards an in-depth analysis of the actors' mind sets and action orientation. The latter is based on qualitative interview analysis and field observation and builds up the core part of the analytical framework. In the following Part 4, the results are discussed by testing the hypotheses followed by a research-based recommendation for an institutional design that could lead the way towards a sustainable development of water management (chapter 8.5).

7. The social-ecological system of Petorca Province

In order to start the empirical research based on the SESF, I chose a focal level as recommended by McGinnis and Ostrom (2014). The idea was to find key indicators, variables or processes that are potentially relevant to the decision-making process of water management in Petorca. For this, the SESF divides each case in its resource system(s) [RS], resource unit(s) [RU], governance system(s) [GS] and actors [A]. These are embedded in social, economic and political settings [S] and related ecosystems [ECO]. Together they influence the focal action situation:

>»In medicine, doctors usually follow a diagnostic approach toward identifying a solution to a medical problem. A doctor will ask us a number of initial questions and do some regular measurements. In light of that information, the doctor proceeds down a medical ontology to ask further and more specific questions (or pre-

scribes tests) until a reasonable hypothesis regarding the source of the problem can be found and supported.« (Basurto et al. 2013:2)

Following this logic and tackling the described research question, it is then important to classify which »of the attributes of a particular SES system are likely to have a major impact on particular patterns of interactions and outcomes« (Basurto et al. 2013:2). This method aims at determining both: factors that are individually important to the case study and factors that are common to comparable cases (ibid.). In this case study, the different sections were determined as follows: To describe the focal situation an analytical scale in terms of geographic and socio-political level was selected. As the water crisis in Petorca concentrates on the river basins of La Ligua River and Petorca River, the surrounding districts Cabildo, Petorca and La Ligua were selected on a geographical level (see Figure 6). Further, these are the biggest districts regarding inhabitants as well as land. Consequently, I chose this level also for describing the social-political system. However, as this focus area is embedded and interrelated to the broader geographical and political context, quantitative and qualitative data of regional, national, and international level were considered if necessary. Hence, the river basins Petorca and La Ligua and their direct surroundings were selected as the relevant resource systems [RS] and water as their resource unit [RU].

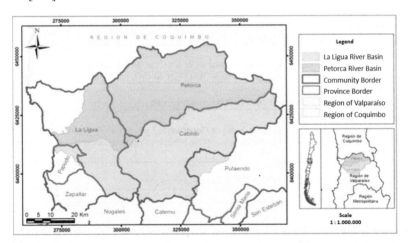

Figure 6: Map of Petorca Province. (Adapted from DGA 2012 in CNR and Universidad de Concepción 2016: 26 modified by Roose)

Relevant actors are those who directly or indirectly influence the decision making in water management. Equally, policy areas that directly or indirectly concern water management and could influence the behavior of these actors (e.g. agriculture, mining etc.). were considered in the presentation of the governance systems [GS]. The sectors interactions [I], referring to interaction between actors, and outcomes [O], referring to social and ecological performance measures and externalities to other SESs, suggested by McGinnis and Ostrom (2014: 5) for describing the action situation are being left out in this study and are not included in the list of tiers. However, by way of the combination of SESF and AI, interactions and outcomes are considered in sector [A].

Each section [RS; RU; GS; A; S; ECO] contains various second and third tier variables. As a next step, research relevant second and third tier variables of each section were chosen. For this, the tiers proposed by McGinnis and Ostrom (2014: 5) were taken as a base and modified according to (a) literature research - especially relying on the studies conducted by del Mar Delgado-Serrano and Ramos (2015) and Basurto et al. (2013), (b) assistance of scholars working in the research area as well as (c) observation of the conflict. The following Table 2 shows the resulting first and second tiers[13]:

Table 2: Categories and tiers. (Based on McGinnis and Ostrom (2014) and Roose)

[S] Social, economic, and political settings [S1] Economic development [S2] Demographic trends [S3] Political stability [S4] Other governance systems [S5] Markets [S6] Technology
[ECO] Related ecosystem [ECO1] Climate Change Patterns [ECO 2] Pollution Patterns
[RS] Resource System [RS1] Sector [RS2] Clarity of system boundaries & location [RS3] Size [RS4] Human constructed facilities [RS5] Productivity of system [RS6] Equilibrium properties
[RU] Resource Unit [RU1] Growth [RU2] Number of units [RU3] Resource value
[GS] Governance System [GS1] Policy Area [GS2] Geographic Scale of Governance System [GS3] Regime type [GS4] Rule making organizations [GS5] Rules in use [GS6] Property-rights systems [GS7] Historical continuity
[A] Actors [A1] Relevant actors [A2] Social unit* [A3] Action orientation*
*Tiers developed by Roose

On this basis, I accumulated a list of qualitative and quantitative variables that specify their definitions, scales, data sources and research tools. The

---

13 The complete list of tiers can be found in the online annex (www.nomos-shop.de/isbn/978-3-8487-6930-8).

information needed for the categories of [S], [RS], [RU], and [GS] is mostly available in secondary resources. Nevertheless, primary sources were taken into account if needed to include local or qualitative knowledge (cf. del Mar Delgado-Serrano and Ramos 2015).To collect qualitative data, I developed a semi-structured questionnaire that mostly targets the data concerning the sub-tiers of [A] actors. The list of qualitative and quantitative second- and third-tiers was modified during the empirical process as insights of the conflict's complexity and the availability of data altered the estimated indicators. As the empirical research of the present study started in 2015 and was mainly conducted in 2016, the data of the following chapters cover mainly this time. However, the available data vary and cover mostly the timeframe from 2013 to 2015. The state-owned online platforms *Biblioteca del Congreso Nacional de Chile* (BCN) and *Instituto National de Estadísticas Chile* (INE, National Institute of Statistics Chile) offered most of the needed quantitative data. Besides this and other minor sources, quantitative data of Petorca were obtained from a publicly available study on the two river basins made by the CNR and Universidad de Concepción (2016) for the Ministry of Agriculture (CNR – Comisión Nacional de Riego, Ministerio de Agricultura) and publications of the National Water Directive (DGA).

The results of the application of the analytical framework lead to the following description of the research case. The chapters zoom in from the broader context shown in 7.1 and 7.2 over a meta-level in chapters 7.3 and 7.4 and finally end in an in-depth actor analysis in 6.5. While the former chapters 7.1 to 7.4 are based on the SESF according to Ostrom (2007, 2009), the last chapter 7.5. follows the analytical framework of AI by Mayntz and Scharpf (1995).

7.1 Social, economic, and political setting

Based on the SEFS, the focal situation of this study is embedded in related ecosystems as well as in a social, economic, and political setting. The following part describes the social, economic and political setting by taking the sub-tiers developed by Ostrom (2014) and del Mar Delgado-Serrano and Ramos (2015) as a guideline.

[S1 ECONOMIC DEVELOPMENT]
According to the data from the INE of 2013 (BCN 2018a, 2018b, 2018c) the strongest sector in all three districts regarding employment is the ter-

tiary sector (service sector). Especially the community of the province capital (La Ligua) has most employments in this sector. The second largest sector is the primary sector (raw materials), consisting mainly of small, medium and large-scale agriculture as well as small and medium mining activity. Regarding water governance, agriculture is the sector with the highest water demand extracting 86 % respectively 77 % of the groundwater from river basins Petorca and La Ligua (CNR and Universidad de Concepción 2016: 232-233). In the Vth region large-scale agriculture dominates the land use holding approximately 90 % of cultivated land while only making up 7,7 % of all water right holders. Small-scale farmers on the other hand only cultivate 4,02 % of the land dedicated to agriculture while making up for 80,7 % of the water right holders (ibid.). Comparing the data of 2009, 2011 and 2013 a shift in the dominant sectors can be noticed. While in 2009 agriculture offered more than twice as much employment as the secondary sector, in 2013 the differences between the two sectors were very slight as agriculture had shrunk drastically. Especially Petorca suffered from job losses. From 2009 to 2013 the agricultural business lost nearly 80 % of its employments. In absolute numbers this sums up to a loss of 920 jobs while the secondary and tertiary sectors together only provided 237 additional jobs (BCN 2018a, 2018b, 2018c). Hence, the other sectors did not compensate for the job losses in agriculture. In the whole province unemployment grew continuously from November 2015 to May 2016. The latest data available show 8 % of unemployment (Gobernación 2016: 5). According to the government of Petorca, the growing increase of unemployment is caused by the dependence of the commercial sector on the extractive sector, especially on agriculture (ibid.).[14]

Unfortunately, detailed data about income per capita and income dispersion is not available on a community or provincial level on the government websites. Nevertheless, a study conducted by Gonzalo Durán and Marco Kremerman of the Fundación Sol accessed the INE data on income dispersion via the governmental program Transparencia Activa[15] and pub-

---

14 It should be noted that these data reflect on formal employment only. It can be assumed that informal employment took a similar development. However, an analysis of the informal employment outreaches the scope of this study.

15 Transparencia Activa obliges governmental bodies to provide certain information like organizational information, contacts and relationships in order to higher transparency. This information is made available via governmental bodies of the program *GobiernoTransparente* (Gobierno de Chile 2018).

lished them in September 2016 (Durán and Kremerman 2016). Duran and Kremerman show that inside the Vth region of Chile Petorca, Quillota and San Felipe are the provinces with most unsatisfactory income dispersion while also being the provinces with the lowest income average (Durán and Kremerman 2016: 7). Additionally, to estimate the inhabitants' financial power, they considered the percentage of people living in income poverty. The data of 2011 and 2013 available on INE show that income poverty is above national and regional average in all three districts. Nevertheless, it decreased between 2011 and 2013, although only slightly in Petorca (BCN 2018a, b, c; see Figure 7)[16]:

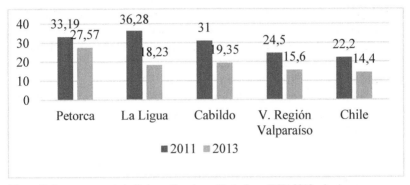

Figure 7: Income poverty in Petorca Province. (Data from BCN 2018a, b, c)

The researchers of the CNR and Universidad de Concepción (2016) compared poverty lines more in detail into »poor« and »extreme poor« measured by income. The data show that extreme poverty is highest in Petorca (8,2 %) compared to the districts La Ligua (5,1 %) and Cabildo (1,3 %). The comparison of the data to the national average demonstrates that the number of persons living in extreme poverty differs only slightly. However, in 2013 the number of persons with general income poverty living in the Petorca Province was 5,7 % higher than the national average (cf. CNR and Universidad de Concepción 2016: 235-236).

---

16   It needs to be mentioned that the indicators for income poverty changed between 2011 and 2014. This means that any changes between those years have limited informative value.

[S2 DEMOGRAPHIC TRENDS]

The Petorca Province is known to be the largest province of the Vth region Valparaiso reaching from the Andean mountain chain to the Pacific Ocean. Due to its low number of inhabitants (75.904) the population density is consequently low as well, consisting of 16,54 inhabitants/km^2 (BCN 2018a, b, c). According to the *Summary of the Provincial Strategy of Petorca* (Resumen Estrategia Provincial de Petorca) published by the Province government in 2016 (Gobernacion 2016), the three districts Petorca, La Ligua and Cabildo are most effected by the water scarcity. Furthermore, they have the highest rural population (La Ligua 22,43 %, Petorca 26,75 % and Cabildo 34,28 %, ibid.). With regard to population growth, age and gender structure the following trends were noted (see Table 3):

- Population growth is lower than the regional and national trend
- Number of old aged inhabitants increases in all three districts and is above national average in Petorca
- Decreasing population between the age of 0-14
- The number of males in relation to females is above national and regional averages but has decreased disproportionately between 2015 and 2017 (compared to national average).

Table 3: Population growth, age and gender structure in Petorca, La Ligua and Cabildo Districts. (Data from BCN 2018a, b, c)

Territory	Population			Index of senior citizens[17]		Index of Masculinity[18]		
	2002	2015	Variation (%)	2002	2015	2002	2015	2017
Petorca	9,440	10,323	9,35	35.36	56.44	103.71	103,17	99.03
La Ligua	31,987	33,878	5,91	31.19	48.85	101.07	102,58	96.07
Cabildo	18,916	20,117	6,35	25.99	40.86	100.17	100,89	95.42
Vth Region	1,539,852	1,825,757	18,57	40.11	62.56	95.66	96,52	No data
Chile	15,116,435	18,006,407	19,12	31.30	50.6	97.12	97,9	95,88

[S3 POLITICAL STABILITY]

The Worldwide Governance Indicators of The World Bank Group (2017) measure »*[...] perceptions of the likelihood of political instability and/or politically-motivated violence, including terrorism*« by several indicators from different sources. Their measurement indicates that Chile's political stability has weakened. While in 2005 it was ranked on the same level as

---

17 Calculated by the number of seniors (65 years or older) in relation to every 100 children (0 to 14 years).
18 Calculated by the number of men in relation to every 100 women.

the other OECD countries at about 73 from 100 points, it has continuously fallen, reaching 59 points in 2015. Nevertheless, compared to other regions of Latin America and the Caribbean, the Chilean political stability is slightly above average (The World Bank Group 2017).

Conflicts is suggested as a indicator for political stability by Mar Delgado-Serrano and Ramos (2015) and is included in the indicator of The World Bank Group. To contextualize the data, conflicts in Chile from 2015 to 2016 will be displayed briefly in the following paragraph. *The Instituto Nacional de Derechos Humanos* (INDH 2012, National Institute for Human Rights) counts 97 socio-environmental conflicts in Chile including the water conflict in Petorca in the year 2012. The type of conflicts referring to water are manifold and vary mostly due to their geographic location and the economic activity related to them; therefore, most water conflicts associated with mining are found in the north, water conflicts connected to agriculture in Central Chile and water conflicts related to energy are mostly present in the south (Valdés-Pineda et al 2014; Rivera et al. 2016). It is not relevant to the purpose of this study to deal with all these conflicts in detail. Nevertheless, conflicts potentially influence the actor's perception of the social and political embedment of their situation. Especially in 2015, several corruption scandals emerged that outraged Chilean citizens and seemed to change their image of their country. For example, a survey by Cooperación Latinobarómetro shows that in 2016, 9,7% of the interviewed Chileans considered corruption as one of the major problems of their country, while in the years before (2009, 2011, 2013) not more than 1,2 % held this opinion (Corporación Latinobarómetro 2018). The Box 1 below shows a few but outstanding conflicts which gained national importance in terms of demonstrations and media cover. Hence, they are relevant for the understanding the political setting of the Petorca conflict for the reader of this study, on the one hand, and, for the perception of the political setting of the actors of the focal situation on the other hand.

Besides the water conflict, there are several conflicts on a local scale investigated in this paper. Starting in 2015 in La Ligua, a number of people organized against a project of a thermoelectric power plant which was planned to be built by a private company in the sector of Quebradilla (La Ligua) (field observations). Citizens expressed their concerns about the project's impacts on health and environment on different occasions such as demonstrations, government reunions etc. In 2017, protests against another already existing industry increased. A turkey production and marketing company runs one of its cattle industries close to the city center of La

Ligua which occasionally causes bad smells all over the urban area due to the animals' excrements. Furthermore, in the Cabildo District present concerns are mainly about air and water pollution due to the mining industry located next to the town center of Cabildo. People claim that there is an abnormally high rate of stomach cancer due to minerals released by the mining company and poisoning their water supply (field observations).

'Red Tide' Catastrophe:
The red tide catastrophe in 2016 was named after the color change that a toxic algae bloom provoked in the South Pacific close to the island Chiloe. This phenomenon caused poisoning of seafood and outreached its natural recurrence in 2016 causing environmental and economic damage, especially in the fishing industry. Fishermen and local citizens blamed the large salmon industry for causing the tide and protested to the government claiming for higher compensation of their losses (Pfeiffer 2016; BBC Mundo 2016; The Guardian 2016)

Corruption allegations against President Bachelet family members:
In 2015 several corruption allegations were raised against the president's son and daughter in law, leading to a prompt and drastic fall of the president's approval rates (84 % in 2010, to 31 % in 2015). Andrónico Luksic, nationally known as of Chile's richest men, was involved in this scandal which induced the Chileans anger about the strong relationship between economy and politics. Several newspapers reported that the corruption allegations against the presidents' families appeared together with other allegations against the opposition and changed the nation's image insofar as Chile had formerly been internationally recognized as a country with low corruption (Romero 2015; Bonnefoy 2016; Franklin 2015).

Toilet Paper Cartel:
In 2015 two large paper manufacturers in Chile were charged for fixing toilet paper prices for over a decade. International newspapers reported that these companies made up for over 90 % of the paper market, and the scheme outraged many Chileans. In addition, one of the companies was owned by Gabriel Ruiz-Tagle, a formal minister of the previous government of president Sebastián Piñera. Hence, the scandal reiterates the image of a strong relationship between politics and economy (Montes 2015; The Guardian 2015).reiterates the image of a strong relationship between politics and economy (Montes 2015; The Guardian 2015)

Box 1: Outstanding conflicts in Chile between 2015-2016

Moving on with the list of sub-tiers, Mar Delgado-Serrano and Ramos (2015) suggest respect for democratic values to measure political stability. Such values are normative, and subjective indicators can be questioned. In this study, I use surveys indicating the preferred political system of Chilean citizen and voter turnout in presidential elections. Unfortunately, there are no data available on a provincial level. Nevertheless, a survey published by Corporación Latinobarómetro (2018) shows that in 2015 only

64,8 % of 1200 interviewed people in Chile regard democracy as the preferable political system. The number has constantly increased since 2007. Nevertheless, in comparison to other states of the Southern Cone, this number is constantly the lowest from the beginning of the survey in 1995 until 2015. According to a study of Pew Research Center (2017), only 5 % or less of the Chileans trust their government (average 14 %), and 24 % would prefer a political system other than democracy. Corporación Latinobarómetro (2018) has a greater number of respondents and confirms that 17,9 % believe that democracy is not the best political system. This rejection towards the political system is reflected in the voter turnout in presidential elections that remains one of the lowest worldwide. Since 2010 it has been constantly decreasing and reached a short rise to comparably low 46 % of participators during presidential election in 2017 (Ríos 2017).

Another sub-tier I would like to stress to indicate political stability in Petorca is norm compliance, represented by the reported crime rates.[19] [20]. It can be noted that in every district the reported crime rate is under national and regional average and it decreased during the years 2010-2014. The lowest crime rate is found in Petorca (see Table 4). It needs to be mentioned that crime rate cannot be taken as a sufficient indicator for norm compliance as mistrust to the justice system or intimidation can trigger the number of undetected cases. Hence, these data are primarily intended to serve as an indicator for crime reports in comparison to the region and the country.

Table 4: Reported crimes per year in Petorca, La Ligua and Cabildo Districts. (BCN 2018a, b, c)

Territory	2010	2011	2012	2013	2014
La Ligua	2,730.41	2,625.64	1,915.69	1,958.52	1,961.83
Petorca	1,393.16	1,557.38	1,454.62	1,279.30	921.85
Cabildo	2,132.35	2,125.23	1,611.83	1,499.43	1,378.99
Vth Region	3,127.05	3,347.34	3,000.45	2,874.96	2,899.35
Chile	2,780.25	3,010.10	2,720.38	2,730.08	2,801.19

19 Number of crimes including different kinds of robbery (violent robbery, car robbery etc.), murder, injury, and rape in relation to every 100 thousand inhabitants.
20 Norm compliance regarding environmental standards and water is discussed in chapter 7.1 and 7.2.

[S4 OTHER GOVERNANCE SYSTEMS OR GOVERNMENT RESOURCE POLICIES]
The sub-tier [S4] should give an overview of the *»top-down policies adopted by the national, regional and local governments to manage natural resource«* (McGinnis and Ostrom 2014; del Mar Delgado-Serrano and Ramos 2015). The data shown concentrate on the inter- and national governance system concerning water, whereas the local governance system is described in chapter 7.4.

The scope of action in Chilean water governance is strongly limited by the National Water Code (Código de Agua) introduced during the military regime of Augusto Pinochet in 1981 (Larraín 2010: 9; Saleth and Dinar 2005: 11; Valdés-Pineda et al. 2014: 2553; Clarvis and Allan 2014: 83). Since then, water rights have been privatized in different user areas, and power structures have shifted from national to private dominance (OECD 2015: 99, Larraín 2010: 13–16; Clarvis and Allan 2014: 83). For example, the 16 public water supply companies were privatized and turned to transnational companies during democracy. Consequently, since 2004 almost all water rights for human consumption in Chile have been in the hands of the private sector (cf. Larraín 2010).

After 15 years of political discussion, a water reform towards more democracy and sustainability was introduced in 2005 (Larraín 2010: 8; Clarvis and Allan 2014: 83). The four supposedly most important steps since 2005 aiming for institutional improvements include: a) the National Strategy for Climate Change (Estrategia Nacional de Cambio Climático, ENCC) introduced in 2006, b) the National Integrated Watershed Strategy (Estrategia Nacional de Gestión Integrada de Cuancas Hidrográficas, ENGICH) introduced in 2007, c) the creation of institutions to serve ENGICH and d) the creation of a Ministry of Environmental Issues (Ministerio de Medio Ambiente) established in 2010 (Larraín 2010: 27; Retamal et al. 2013: 12). Retamal et al. (2013: 12) state that these steps have introduced an interconnected conception of socio-ecological policy regarding water as a development driver which paves the way for climate change mitigation and adaption. Furthermore, in 2009 Chile signed a ministerial declaration on water management at the Vth World Water Forum in Istanbul in which it recognizes water and sanitation as a human right and agrees to work towards the implementation of this declaration in national policy (Larraín 2010: 27).

The criticism of »compliance of the environmental regulatory and policy frameworks« (sub-tier of S4) concerning water management is manifold. The OECD criticizes the Superintendence for the Environment (Superintendencia del Medio Ambiente de Chile, SMA) that oversees the organization and coordination of monitoring and control of all environmental issues established by law (including environmental qualification resolutions, prevention plan and environmental decontamination measures, the composition of environmental quality standards and emission standards, as well as management plans) for being incapable of acting effectively on the local level. Especially, the OECD complains about too little data on the »*environmental impacts generated by the medium- and small scale mining industry*« on land and water contamination (OECD/ECLAC 2016: 28-29). The OECD underlines that Chile is the only member country which has not introduced criminal penalties for egregious environmental offences and recommends doing so. Moreover, the OECD endorses efforts that lead to stricter rules, transparent data and increased monitoring such as:

> »● Introduce a strict liability regime for future damage to water bodies, land, species and ecosystems; develop and implement environmental remediation standards and plans, particularly for soil contamination with heavy metals; empower the SMA to enforce liability provisions through administrative actions.
>
> ● Maintain and frequently update risk assessment inventories of abandoned contaminated sites; establish a financial instrument to impose decontamination fees on hazardous industrial installations and mines and earmark the revenue to constitute a fund to be used for clean-up of past land and water pollution.
>
> ● Upscale efforts to monitor and control the resource use (e.g. energy, water) and environmental impacts generated by medium- and small-scale mining industries, and support the adoption of new processes and technology to increase their efficiency and safety.« (OECD/ECLAC 2016: 29)

As part of the OECD, Chile has been encouraged to meet their standards in water management. The OECD reports state that there has been an improvement since 2005 (OECD 2005, OECD/ECLAC 2016). Nevertheless, current OECD recommendations demand the development of measurement and monitoring strategies of water quality by setting quality standards for all rivers and groundwater bodies and the improvement of water management by prioritizing human consumption and sanitation, strengthen market transparency and set effective and enforceable abstraction limits to stay within ecological and social limits (OECD/ECLAC 2016: 78).

[S5 MARKETS]
Chile's most prosperous economic activity is the export of raw materials, for example by the mining industry (extraction of copper), agriculture industry, forest industry and salmon industry. Hence, the intended market is foreign. The second important sector encouraging economic growth is the domestic sector. The financial business relying on transaction etc. prospers

as credits are needed in almost all issues of daily life due to privatized education, health and pension system, among others (Benedikter and Siepmann 2015: 20–21). As far as the Petorca Province and the water conflict are concerned, there are three economic activities with higher influence on the social, political and economic setting to be mentioned: (a) water rights (b) agriculture (c) mining.

To a) type of product: (water) rights, mainly underground water rights, *influence of global / local market*: water rights are owned by transnationals and locals that influence prices; further it should be mentioned that an informal water market was identified by researchers of the CNR and Universidad de Concepcion referring to local water rights owners that sell their rights to the local water provider ESVAL S.A., the municipality or local water associations (CNR and Universidad de Concepción 2016: 379ff). *Access to market:* barriers of education and financial power, legal barriers in case of not consistently registered water rights, *market incentives*: fee for non-use, *demand for natural resource*: water

To b) type of product: agricultural products (mainly vegetables & fruits), *influence of global market:* demand from Europe and United States especially for avocado, increases, exported volume between 2006 and 2016 increased by 32,7% while the exported value increased by 216,5% (ODEPA 2017: 68). Therefore, it influences local land use and crops such as plantations of avocados and citric, *influence of local market:* demand of local grocery stores and farmers markets might influence crops, *access to market:* international market requires certain amount of goods to enter under reasonable costs, *market incentives*: governmental subsidies for large-scale agricultural companies, governmental consultancy for medium and small scale farmers, *demand for natural resources*: water and land

To c) *type of product:* raw material (copper, gold, silver) *influence of global market:* mining companies depend on international prices for copper and other minerals. The copper prices affect the operational costs. For example, when copper and mineral prices go down, the operational costs increase the pressure on mining companies, *demand for natural resources*: minerals, water

[S6 TECHNOLOGIES]
The sub-tier [S6] is included in the list of indicators »as a potential source of exogenous shocks, and generalization of market incentives to any factors relating to markets and government resource policies to other potentially relevant governance systems.« (McGinnis Ostrom 2014: 7). For the present research case the following technologies need to be mentioned be-

cause they are potential drivers and/or solutions for the water conflict and will be examined later with regard to the hypotheses (cf. chapter 8). Currently, the technology commonly used to cope with drought and improve water efficiency depends on technical irrigation systems and tank reservoirs, in addition to the traditional infrastructure, such as wells and channels. In Chilean agriculture professional technical irrigation systems are increasingly used. Following law 18.450, the Chilean government subsidizes small and medium scale farmers in improving their irrigation systems. Tank reservoirs are storage spaces for fluids. In the Petorca Province several large-scale water tanks were built by large-scale agricultural companies to store rainfall for irrigation purposes. These private tank reservoirs can be subsidized by the national institute CNR (cf. CNR and Universidad de Concepción 2016: 227). Moreover, the Institute of Farming Development (INDAP) offers a program to build shared tank reservoirs for small-scale farmers organized in WUOs. For instance, in Petorca INDAP subsidized a water tank of the WUO La Viña-La Vega in 2017 which can store water for the use of 25 small-scale farmers who cultivate a total of about 40 h of land (INDAP 2017). Moreover, the Ministry of Public Works (MOP) plans to install two large scale tank reservoirs in the province which are supposed to provide water for both, big and small-scale agriculture and human consumption. The water reservoir Las Palmas in the Petorca District containing a volume of 55 million m^3 of storage capacity will primarily serve irrigation purposes (Ministerio de Obras Públicas 2018a). In times of deficiency, this water will also be used to supply drinking water in rural areas (to support the APR). To build this reservoir, 252 ha need to be flooded. The project was tendered in July 2017. The second large-scale tank reservoir project, called Los Àngeles, is supposed to have a storage capacity of 30 hm^3 and will be installed in the La Ligua District (Ministerio de Obras Públicas 2018b). The tender was supposed to open in the first semester of 2018. Further water reservoir projects of the MOP are the tank Pedernal/Sobrante and the tank La Chupalla. Both projects are in their initial planning phases, for which no data could be found.

Considering upcoming technologies, two mainly discussed emerging solutions should be mentioned: *desalination plants* and '*water highway*' (south-north water pipeline). The former is currently used by the mining industry in the northern parts of Chile. However, this technology could potentially provide drinking water as well. In the La Ligua District a desalination plant prototype exists. This project is strongly supported by the mayor. The idea behind it is to find a technical solution to the water scarcity by creating greater supply. However, this project has often been criticized, mainly for being too cost-intensive (cf. Interviews 1, 4, 5, 19). Yet, a regulative framework for the use of desalinated water does not exist. The so-called *water highway* describes the idea of transporting water from the rivers of the humid south to the dry north of Chile. However, there were no data found indicating plans of actual implementation. Nevertheless, members of the Petorcean large-scale agriculture association (here called LSAA) have highlighted this idea as a long-term solution for the future (Interviews 24, 25).

7.2 Related ecosystems

The focal socio-ecological system of this study is the water system of Petorca Province, geographically referring to the two river basins of Petorca River and La Ligua River. The social, economic, and political setting frames the focal socio-ecological system of this study. Likewise, related ecosystems [ECO] frame the research case. Taking the sub-tiers of the section [ECO related ecosystems] suggested by Ostrom (2009) [ECO1 climate patterns] and [ECO2 pollution patters] as a base,[21] the following description mainly concentrates on the climate patterns influencing the Petorca water system. For this study, understanding those patterns is important to estimate the ecological reasons of the water scarcity and the interrelation with anthropogenic interventions. It shows that Petorca has been hit by a severe and ongoing drought since 2010. Although the region is affected by the El Niño-Southern Oscillation (ENSO), data indicate that the current drought is independent of that effect and seems to be caused by anthropogenic climate change (Garreaud et al 2017). Standards for norm compliance assurance for water quality in Chile are rather low. As there are few data available on pollution patterns, this sub-tier will be discussed only briefly. In short, the norm of water quality seems to be complied with. Nevertheless, mining and agriculture seem to provoke a rise of pollutants.

[ECO1 CLIMATE PATTERNS]
As Chile is characterized by the high diversity of its climate zones, it is not appropriate to identify national climate trends. For the same reason,

---

21   Ostrom (2009) suggests a third sub-tier: [ECO3 Flows into and out of focal SES]. However, in this study those flows are described in [RS6 Equilibrium properties].

the Fifth Assessment Report of the Intergovernmental Panel on Climate Change (IPCC) shows different outcomes of climate modelling relevant to Chile, depending on the geographical context. Increased temperatures are likely to occur in the north; while less intense temperatures are likely to occur towards the south. These trends have already been visible by the warming in the Central Valley and Andes mountains over the last decade (IPCC 2015). For this study, it is reasonable to focus on the sub-national level of central Chile. Petorca (approx. S32°00 – S32°41) lies in the northern part of central Chile (S30-38°) and is marked by the Mediterranean-like, semi-arid climate of that area (Garreaud et al. 2017: 6310; MOP and DGA 2015: 12). According to the National Water Directive (MOP and DGA 2015), the average annual rainfall of central Chile (here: S32°-36,5°) is 943 mm/year and is basically concentrated on the winter months (April-August). The dry season lasts for about 7-8 months. The northern part represented by the region of Valparaíso (including Petorca) is the driest, characterised with an average rainfall of 434 mm/year (MOP and DGA 2015). In addition, El Niño-Southern Oscillation (ENSO) influences precipitation in central Chile. During the years of ENSO low pressure usually contributes to more rains during the winter months, whereas during La Niña high pressure on the east side of the Pacific Ocean intensifies. As a consequence, precipitation decreases (cf. Garreaud 2017 et al.: 6311). The La Niña phenomenon happens irregularly, approximately every 2 to 8 years. During the time between 2007 and 2018, the strongest La Niña happened in the years 2007-2008 and 2010-2011. The past weak La Niña years were between 2008 and 2009, 2011 and 2012, 2016 and 2017 as well as 2017 and 2018 (Australian Government 2018).

Garreaud et al. (2017: 6311) conducted an intensive study on climate patterns in central Chile focusing on regional-scale meteorological dry spells. Considering 25% of annual rainfall deficit as a drought event, they found that between the years 1916 to 2009, 19 of such events happened randomly, mostly every 2 to 9 years (one 14 years pause between 1924 and 1938). Most events lasted one year or two; two events reached 3-year duration (1967-69 and 1994-94). From 2010 to 2015 they found a four years long period, which is unique in the analyzed time frame. Therefore, they called this event a mega-drought. The mega-drought is characterized by its spatial scale as it reaches further south than the previous drought events (cf. ibid. 6313). In addition, according to Garraud's et al. study, using tree-ring-based rainfall reconstruction, the last decade (2004-2014) including the years of the mega drought, report extraordinarily low precipitation which exceeds the millennium context with only few possible comparable phenomena (cf. ibid. 6316). Garreaud et al. compared their data on the mega-drought to the ENSO data (2009 – 2015). As a matter of fact, the drought took place mostly during ENSO-neutral years, once during La Niña (2010) and once during El Niño (2015). Hence, the data indicate that other factors than the ENSO provoked the drought event. Garreaud et al. state that their outcome is in consonance with the work of Boisier et al. (2016) as they »*[...] estimate that as much as a quarter of the rainfall deficit during the MD [mega-drought] is attributable to anthropogenic climate change, mediated by altered mid- to high-latitude circulation in the Southern Hemisphere*« (Garreaud et al. 2017: 6317). According to a study of the CNR and Universidad de Concepción (2016) and based on data from 2015, precipitation rates in Petorca have been fluctuating but remained constantly low since 2008 and 2010 respectively.

Besides rainfall, glaciers play a significant role in the Chilean water system as they melt in the summer and provide fresh water to the rivers and recharge groundwater aquifers. Studies show that nearly all glaciers in Chile are decreasing due to increased temperatures and decreased participation (Rivera et al. 2002; Valdés-Pineda et al. 2014: 2559). In central Chile especially, the situation has reached such an extent that towards the end of the hottest season almost no snow reservoirs could feed rivers or groundwater aquifers (cf. Valdés-Pineda et al. 2014: 2557; Garreaud et al. 2017: 6311). In Petorca, there are no glaciers feeding the rivers Petorca or La Ligua as the existing ones melt into the flow of the Aconcagua River (MOP and DGA 2015:13). However, there are no data available indicating in how far the glaciers feed into the groundwater system. Prognostics for the future development are pessimistic about any relief of the water availability because of decreasing glaciers and precipitation. This situation increases the pressure to develop sustainable water governance that copes with these hydrological uncertainties (cf. Valdés-Pineda et al. 2014: 2557, 2563; Henríquez et al. 2016: 36).

Furthermore, Garreaud et al. (2017) describe several climate phenomena that happened in central Chile during the mega drought and that could be relevant to the present case of Petorca. Although it may seem trivial, in terms of water scarcity it is in fact crucial that river stream flow declined due to rainfall deficit and snowpack reduction. Garreaud's et. al. findings on surface hydrology indicate that the mega drought caused a lower stream flow than in previous dry spells in the northern part of central Chile (north of 33°S including Petorca), making it »*less productive*« and »*suggesting changes in the runoff generation processes*« (Garreaud et al. 2017: 6323–6324). These findings coincide with the studies of Valdés-Pineda et al. (2014: 2560). They indicate that such changes in river flow and seasonality need to be recognized in infrastructure planning and will influence economic activities. Also, the results overlap with the data on stream flow of Petorca and La Ligua. The DGA (MOP and DGA 2015) compared the yearly mean stream flow to the average stream flow of the years 2013-2014 detecting a variation of - 82% and - 86% respectively. This is the highest reduction compared to the analyzed rivers of Chile except of Rio Huasco (III Region -84%). Furthermore, Garreaud et al. state that »*The volume of water stored in different reservoirs and hydrological systems also dropped dramatically during the MD [mega drought].*« (Garreaud et al 2017: 6318) However, there is not enough information available to estimate the temporal and spatial variability of groundwater or its relation to the drought. Nevertheless, it is reasonable to assume that under the conditions of precipitation deficit and decrease of snowpack, pumping rates will rise and increase pressure on groundwater (cf. Garreaud et al 2017: 6321; Valdés-Pineda et al. 2014: 2563).

In the context of the drought, Garreaud et al. (2017) found that while the average maximum temperatures have increased since 1970, the average minimum temperatures are not showing any notable trend, »*thus causing an increase in daily average temperature and its diurnal range*« favoring potential evapotranspiration from water bodies and vegetation. Especially in the central valleys, potential evapotranspiration »*increased more than 50mmyear−1 during the MD [mega drought]: the interior valleys of northern Chile (30–33°S) and to the south of 36°S, suggesting substantial water stress for vegetation in these areas.*« (Garreaud et al. 2017: 6319) Following this, the researchers also investigated changes of vegetation productivity during the mega drought. According to their data, forest plantations and croplands were not negatively affected by the mega drought. However, severe browning was observed in the northern region ($\leq 33°S$) and at the coastal area in the south ($\geq 39°S$):

> »The browning in the north is substantial, up to −20% of the historical mean at the grid level, and coincides with the region of MD-averaged rainfall deficit in excess of 30% [...] and a PET [potential evapotranspiration] increase over 100mm [...].« (Garreaud et al. 2017: 6320–6321)

In conclusion, although data may differ, the common refrain that is heard in discussions on the impact of climate patterns on water resource is that there is an increasing pressure on water availability in central Chile. As Petorca Province is in the northern part of central Chile, it seems to be especially vulnerable due to outstanding precipitation deficit, increasing temperatures, and the fact that the valley is not fed by glaciers. Consequently, the pressure on groundwater in the Petorca Province might rise. The raise of evapotranspiration might increase the water demand for irrigation. Further, data suggest that this increased vulnerability is likely to be caused by anthropogenic climate change (Garreaud et al. 2017; Valdés-Pineda et al. 2014: 2563)

[ECO2 POLLUTION PATTERNS]
Pollution of the water system can be influenced by environmental factors such as vegetation, topographic or geologic factors and/or anthropogenic factors (cf. CNR and Universidad de Concepción 2016). Regarding the latter, the OECD identifies three major driving factors for surface water pollution in Chile:

> »[1] urban and industrial wastewater, [2] fish farming and processing, [3] and the agriculture and agro-food industry [...], with substantial regional variations. Limited tertiary wastewater treatment [...] and large agricultural runoff have resulted in nutrients contamination and eutrophication of coastal lakes, wetlands and estuaries; mining effluents have increased the concentration levels of heavy metals and other toxic pollutants in surface waters [...].« (OECD/ECLAC 2016: 25)

In addition, ground water pollution is also possible, especially with regard to agriculture and the use of fertilizers or irrigation with sewage (cf. Valdés-Pineda et al. 2014: 2562–2563). The OECD/ECLAC (2016) recommends Chile to improve measurements and monitoring of water quality. In specific, the organization demands the implementing of quality standards for all rivers and quality standards for ground water. Hence, there is little data availability to estimate water quality and pollution patterns. However, a report on the river basins La Ligua and Petorca made by the CNR and Universidad de Concepción (2016) asserts high water quality. Nevertheless, analyzing data provided by the DGA from 2011 to 2015, the researchers of the Universidad de Concepción found patterns that indicate pollution from agriculture and mining industry. In addition to other kinds of pollution, the studies determine an above norm occurrence of mercury (possible sources: industrial release, WHO 2005) and boron (possible sources: industrial production release, pesticides and agricultural fertilizers, WHO 2003) Further, in some tests the maximum norm level of pH was established. Also, a high concentration of different chemical elements from agricultural fertilizers were found. However, the data did not exceed the existing norms (CNR and Universidad de Concepción 2016: 252ff). In summary, the data on pollution patterns show that in Chile and especially in Petorca water systems are threatened by pollution of agriculture and mining industries as well as domestic wastewater. Although, data indicate norm compliance, it should be mentioned that the Andean state has been criticized for its low compliance assurance on water quality norms (cf. OECD/ECLAC 2016: 108; chapter 7.1).

7.3 Resource system and resource unit

This chapter displays the characteristics of the resource system of the two main river basins in Petorca: La Ligua and Petorca and their corresponding ground water systems. Referring to the SESF by Ostrom (2007, 2009), the following indicators reflect the ecological part of the system including the resource units. However, the framework aims at reflecting the interdependencies of the different issues. Therefore, this chapter needs to be understood as to be embedded in the chapters described before on social, economic, and political settings [S] and related ecosystems [ECO] (cf. chapters 7.1 and 7.2). Further, the data presented often refer to the governance system and its actors, which are described in detail in the following chapters (cf. chapters 7.4 and 7.5).

[RS1 SECTOR], [RS2 CLARITY OF SYSTEM BOUNDARIES AND LOCATION] AND [RS3 SIZE]
The river basins La Ligua and Petorca and their ground water aquifers form the resource systems natural boundaries for this study. Both are in the northern part of the region of Valparaíso between 32° and 32`40°S, and their drain flows NE-SW (cf. CNR and Universidad de Concepción 2016: 242). The rivers Petorca and La Ligua have their source in the Andean foothills and drain approximately 1.986 km² and 1.980km² reaching a length of 76 km and 106 km respectively. Before flowing into the Pacific Ocean, the rivers connect when the Petorca River empties into the La Ligua River at the La Ligua bay (CNR and Universidad de Concepción 2016: 220ff). As seen on the image below, the river basins cross over the political boundaries of the districts Petorca and La Ligua (River Basin of Petorca) and Cabildo, northwest of Putaendo (politically belonging to the Province of San Felipe), La Ligua, and northeast of Papudo (River Basin of La Ligua).

Following a study of the river basins made by the University de Concepción (2016: 229-230), the groundwater aquifers are closely linked to the rivers as they depend on their surface water availability. This leads to a situation of limited stability as the groundwater availability responds rapidly to water flow changes in the river. Further, the scientists point out that the groundwater system is closely connected before the rivers reach the Pacific Ocean rivers downstream from the districts Longotoma (River Petorca) and La Ligua (La Ligua River) respectively.

The shape of the surface river flow has not been intervened by anthropogenic processes. Following Chilean law, rivers, as well as lakes and beaches, are national goods of public use and must be accessible for everyone (Código Civil, Art. 589, Art. 595). Nevertheless, some [RS2b anthropogenic boundaries] were reported informally to have been enclosed by fences that impede free access to the rivers. The property of land and water is separated by Chilean law and the river water is privatized (Constitution, Art. 19. Nr. 24; Water Code Art. 5). Both rivers end in the Pacific Ocean and in contrast to many other Chilean rivers, they both do not originate in the Andean Mountain Range but close to it (see Table 5). This is important because it means that they are not fed by the glaciers.

Table 5: Basic data about La Ligua River and Petorca River. (DGA 2018)

River	Basin surface (km²)	Length (km)	Average Annual Flow [m³/s]	Dried out
Petorca	1,988	79	1.1	1997
La Ligua	1,980	80	1.4	2004

The [RS4 ANTHROPOGENIC STRUCTURES FACILITATING RESOURCE MANAGEMENT] such as human built water infrastructure in Petorca are manifold. In the following, I will mention only technical interventions in the resource system. The agriculture industry uses canals, technical irrigation system, wells (formal and informal), drains (formal and informal) and water reservoirs. The latter are mainly used for medium and large-scale farming activities and rural potable water supply. However, due to the ongoing water scarcity, further water tanks for irrigation, as well as for drinking water supply, have been planned.

In this study, the [RS5 PRODUCTIVITY OF SYSTEM] is measured by stream flow. Thus, the productivity is low as both rivers have been declared to be dried out. For both water bodies, in 2014 a flow of 1984 l/s was detected (Petorca River and La Ligua River, DGA 2014 in MOP and DGA 2015). The study of Garreaud et al. show that the streamflow decreased in the northern central valleys so that »*basins became »less productive*« *during the MD [mega drought] compared to previous dry periods, suggestions changes in the runoff generation process*« (Garreaud et al. 2017: 6318).

[RS6 EQUILIBRIUM PROPERTIES]

The surface water of the river basins depends mostly on precipitation. Because of the basins' topography, they are primarily fed by precipitation and not by snow melting from the Andean Mountains (cf. MOP and DGA 2015: 56). Similarly, its groundwater system is estimated to have low stability as it depends on the surface water flows (cf. CNR and Universidad de Concepción 2016: 229-230). Hence, the resource system is vulnerable to droughts and the desertification of the north. Moreover, both basins are officially declared to be restricted areas due to the ongoing drought. Because of being declared »restricted areas«, infinite water rights can no longer be handed out. However, external anthropogenic impacts on the resource system, such as high exploitation by industrial agriculture, are constantly threatening the equilibrium. In Chile, about 82 % (MOP and DGA 2015) of water is used for farming.

In the Vth Region, the demand of the farming business is estimated around 78 % (MOP and DGA 2015). A study on the water system in Petorca made by the DGA in 2012 shows that surface water is only used in the high zones of the Petorca basin (close to the source). This means that water flowing out of the system for industrial use or human consumption mainly originates from groundwater aquifers. Here, 86 % of the water extracted from the Petorca basin is used 86 % for agriculture and 7 % for potable water, while in the La Ligua basin 77 % of the water is used for agriculture and 15 % as potable water. What aggrvates the situation is the fact that around 17 % of agricultural land belonging to the river basin Petorca (2.070 ha) and around 6 % of agricultural land belonging to the river basin in La Ligua (1.010 ha) are situated on steep hills, which increases the pressure on water (DGA 2012 in CNR and Universidad de Concepción 2016: 232). Although the data could be considered overrated, coming from 2012, they indicate the predominant presence of agricultural water use in both basins (see Figures 8 and 9). In summary, mostly agricultural activity disturbs the recourse equilibrium. It diminishes water availability and alters the flora and fauna of correlating ecosystems (cf. CNR and Universidad de Concepción 2016: 287).

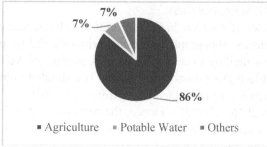

Figure 8: Groundwater use Petorca River basin. (Own illustration derieved from data from DGA 2012 in CNR and Universidad de Concepción: 232)

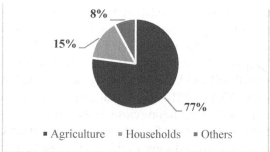

Figure 9: Groundwater use in La Ligua river basin (Own illustration derieved from data from DGA 2012 in CNR and Universidad de Concepción 2016: 232

[RU RESOURCE UNITS] [RU1 GROWTH OR REPLACEMENT RATE] AND [RU2 NUMBER OF UNITS]

The resource units dealt with in the present study are surface and ground water which belong to the mobile and countable resources (del Mar Delgado-Serrano and Ramos 2015: 814).

The growth or replacement rate is represented by river flow development and ground water availability development. Both depend highly on precipitation. Due to the lack of precipitation associated with anthropogenic climate change, growth rate has shrunk constantly, leading to the official drying of the river basins in 1997 for the Petorca River and 2004 for the La Ligua River. Nevertheless, in the winter months between July and November surface water availability rises because of rainfalls (DGA 2013 in CNR and Universidad de Concepción 2016: 245). However, studies of the DGA indicate an ongoing shrinking of river flow (DGA 2014 in MOP and DGA 2015: 62). Studies on groundwater availability between November 2012 and October 2013 by the DGA demonstrate a 255 l/s deficit in the aquifers of Petorca and a 519 l/s deficit in the aquifers of La Ligua between extraction rates and considered sustainable volume (cf. Table 6, CNR and Universidad de Concepción 2016: 248; MOP and DGA 2015: 116).

Table 6: Water rights and water availability in Petorca Province. (Data from MOP and DGA 2015)

Basin	Surface rights left (to the DGA)	Surface rights handed out	Groundwater rights left (to the DGA)	Groundwater rights handed out
Ligua	47	755	1,300	9,209
Petorca	89	3,055	1,312	5,065
Basin	Availability of surface water		Availability of groundwater	
	Permanent	Occasional (m3/s)	Sustainable Volume* [m3/a]	Available Volume** [m3/a]
Ligua	None	1.6 Jul, 2.1 Ago, 1.2 Oct, 1.5 Nov	21,192,192	42,384,384
Petorca	None	2.3 Oct, 2.5 Nov,	11,510,640	23,021,280

*Sustainable volume: annual amount of water associated with the aquifer re-charge. The sustainable volume is constituted as water rights with perma-nent character. Provisional volume refers to the annual amount of water associated to the water rights for underground water with provisional character in areas of restriction. (MOP and DGA 2015: 116).

**Available volume: refers to the sum of sustainable volume in a common catchment area and which can be granted as permanent respectively provisional. (MOP and DGA 2015: 116)

[RU3 RESOURCE VALUE]

The description of resource values, and especially »*values non-recognized by the markets*« (del Mar Delgado-Serrano and Ramos 2015: 819) is normative. A deep analysis of the resource value would extend beyond the range of this study. For this reason, the resource units' value will be contextualized in Petorca in the following - not to quantify its value but rather to display its diverse and normative range of values. Some may argue that water prices could be taken as an indicator for market values as water is privatized in Chile. However, the market value varies considerably. For example, scholars of the CNR and Universidad de Concepción analyzed water rights trade of the two basins between 2001 and 2013. They found a range between the lowest price (sold) of 3,84 UF per l/s ($\approx$ 319 $ as of 9.11.2015) in 2001 and the highest price (sold) of 841,34 UF per l/s ($\approx$ 69.921 $ as of 9.11.2015). Among other factors, prices seem to depend on the amount of water existing in the river and the situation of the vendor because water rights can be sold without physical water availability at the moment of trade. Rights owners tend to sell at higher prices in times of scarcity. The lowest-priced sales were made in the years 2002 and 2001. Compared to the data of Garreaud et al. (2017: 6311), the latest drought so far had been in 1998, three, respectively four years, before the lowest-priced sales were made. The highest sales were achieved in the years 2008 and 2009 - just one respectively two years after the last drought compared to Garreaud et al.(ibid.), as well as in 2011 and 2012. The two last years were squarely in the phase of the mega-drought (ibid.).

However, water is also of cultural, environmental, and strategical value. The latter – in this study described as sub-tier strategic value - helps to point out the political value of water. Water scarcity can be used as a pressuring argument for political goals. In that sense, the water scarcity in Petorca was used as part of campaigns during elections (mayors, governors) and by activists (e.g. campaign for the mayor elections of Luis Soto 'Lucho por el agua', #secos). Experts fear that an alleviation of drought would slow the progress of a new Water Code reform (Interview 28). The Chilean anthropologist Paola Bolados (2016) adds another perspective to the strategic value and claims that the appropriation of water by big private companies is a strategic geographical expansion of capitalism. The sub-tier environmental value is hardly measurable. It is common sense that water is the base for any life on earth. However, its environmental value cannot be quantified. Similarly, the sub-tier cultural value is a qualitative indicator. It is hardly possible to simplify the high normative character into a list of certain general preferences. One the one hand, the Petorca economy (agriculture and mining) depends heavily on water. On the other hand, water is seen as a necessity for a life of dignity, tradition and for recreation. For example, citizens measure in how far live conditions have changed by means of the water scarcity. They report that traditional small-scale farming has decreased and activities like taking a bath in the river and traditional lifestyles like growing up in an environment of greater biodiversity are lost (Interviews 3, 4, 7, 11). In how far different actor groups value and depend on water is analyzed in the sub-chapters of 7.5.

7.4 Governance systems

This subchapter describes the governance systems relevant to the study case. The governance systems are embedded in the social, economic, and political setting as well as in the related ecosystems. Moreover, they interact with the resource system of the river basins Petorca and La Ligua and their resource units: surface and ground water. Further, it sets the frame of possible action situations for the actors described in the following chapters (cf. McGinnis and Ostrom 2014: 7).

[GS 1 POLICY AREA] AND [GS2 GEOGRAPHIC SCALE OF GOVERNANCE SYSTEM]

The relevant policy areas for the focal situation of water governance in Petorca are basically environment, more specifically water management, and agriculture and economy. As these are influenced by different administrative and geographical levels, I focus on a local level but also consider national and international scales if necessary, to explain the observed behavior. It is crucial to understand that the subchapter *governance systems* explains the rules and responsibilities of different actors, while the following subchapter *actors and action orientation* focuses on the action taken by different actors or actor groups (cf. McGinnis and Ostrom 2014: 8).

[GS4 REGIME TYPE]

Chile's government is a representative democratic republic with the president serving as the head of both state and government. The Bertelsmann Transformation Index (BTI) categorizes Chile as a consolidating democracy (Thiery 2016: 4; Merkel 2010: 238). While Wolfgang Merkel, German expert for comparative politics and democratization, classified Chile as a tutelary democracy in 2004, criticizing the BTI by claiming that electoral democracy is not a sufficient indicator (Merkel 2004: 51); he later in 2010 also described the Chilean progress as positive in view of a gradual constitutional consolidation - especially with regard to the constitutional reform of 2005, which stands for the elimination of the authorization enclaves that were inherited from the military dictatorship of Augusto Pinochet (Merkel 2010: 237).[22]

[GS5 RULE MAKING ORGANIZATIONS] in that governance system are all organizations that are »*responsible for crafting and/or implementing different kinds of rules.*« (del Mar Delgado-Serrano and Ramos 2015: 814). Tables 7 and 8 give an overview about the scale and type of organizations relevant to water management in Petorca. It is based on the list of sub-tiers developed by Ostrom et al. (2007) and modified by del Mar Delgado-Serrano and Ramos in 2015 and on the list of actors related to water scar-

---

22   Tutelary democracy is used as a classification for regimes with free and fair elections where the government might be undermined by nondemocratic control from actors that have not been elected. (cf. Kanol 2015: 65)

city identified by the Chilean research institute (CR)^2 (Aldunce et al. 2015)[23].

Table 7: Rule-making organizations. Public Sector. (Aldunce et al. 2015 modified by Roose)

Groups			Level			
			National: Chile	**Regional:** Vth Region	**Provincial:** Petorca	**Communal:** Petorca, Cabildo, La Ligua
Public Sector	Executive		**Ministries:** MINAGRI, MMA, MOP	**Regional Secretaries:** MINAGRI, MMA, MOP	Gobernación **Offices:** Agricultura, Medio Ambiente	Municipalities
			Public Services: CNR, ONEMI, INDAP, CONAF, DGA, SISS, CONADI, SUBDERE, FIA, DOH	**Public Services:** CNR, ONEMI; INDAP, CONAF, DGA, SISS	**Public Services:** INDAP	**Public Services:** INDAP/PRODESAL
	Legislative		**Senate:** Comisión Especial sobre Recursos Hídricos **Cámera de Diputados:** Comisión de Recursos Hídricos y Desertificación	Diputados Districto 10		
	Juridical		Supreme Court	Tribunales de primera instancia Corte de apelación	Tribunales de primera instancia	Juzgado de Letras

The list shows that actors belonging to the public sector predominate (approximately 23 governmental bodies[24]). Scholars criticize the high diversity of public water institutions in Chile (Valdés-Pineda et al. 2014: 2553). Retamal et al. (2013: 13) identify more than 100 organizations that influence water governance in Chile, either directly or indirectly. The current

---

23   The subtiers developed by Ostrom (2009) are: Private sector organizations (for profit), Nongovenrmental, non-profit organizations, Community based organizations and hybrid organizations.

24   The number is considered as approximately because different governmental bodies named here are sub-bodies of others. However, they have different responsibilities and are therefore counted for themselves. Further, it is not excluded that there exist more governmental bodies that have not been counted here.

situation shows a tendency to overlapping responsibilities, competition, and complex bureaucratic regulations. At the same time, the institution responsible for handing out and monitoring water rights (DGA), declares itself as incapable of fulfilling its tasks aquatically because of insufficient human resources and legal power (Diario El Divisadero 2017; Interview 35).

Table 8: Rule-making organizations. Private Sector and Civil Society. (Aldunce et al. 2015 modified by Roose)

Groups		Level			
		National: Chile	**Regional:** Vth Region	**Provincial:** Petorca	**Communal:** Petorca, Cabildo, La Ligua
Private Sector			ESVAL	Agriculture Companies; Large-scale agriculture asociation AC, CASF; CAST	Small-scale farmers AC, CASF; CAST Mining Companies
Civil Society	NGO, Non-Profit	Observatorio Ciudadano; Chile Sustentable; Fundación Newenko; Coordinadora por la defensa del agua y la vida; Fundación Heinrich Böll etc.			APR
	Community-based movements	MODATIMA	MODATIMA	MODATIMA	Colectivo Chasky Minga por el Agua
	Hybrid	Academic Organizations (CR2, CRHIAM etc.)		Programa Gestión Hídrica	Oficina de Asuntos Hídricos Autoridades clericales

[GS5 RULES IN USE]
In order to understand the rules that frame the actors' behavior, Ostrom (1999: 65-72) suggests distinguishing between three levels of rules: (1) Constitutional-choice rules that decide who can (officially) participate in a regulative framework and that set the rules how to build such a framework. (2) Collective-choice rules that define rules for civil servants or other representatives (e.g. WUO director) and (3) operational choice rules

that define the rules for daily decisions and direct management, such as when, who and how to extract resources as well as how to monitor and sanction water use. The different levels of rules can be influenced by formal and informal decision-making processes and influence each other directly or indirectly.

The constitutional-choice rules in Chile are as follows: The Chilean Constitution assures the private persons property of water in its Artículo 19 N° 23 y 24 inciso final, of the *Constitución Política de la República* (Constitution). Further articles concerning water management are: Artículo 5, inciso 1° and 2°, *Constitución Política de la República de Chile, Reforma* 1989, stating that the sovereignty is limited to the respect of basic human needs and that the state is responsible for assuring those rights which are guaranteed by the constitution as well as international agreements. In terms of international agreement, Chile did not ratify the Protocol of San Salvador that assures the right to a healthy environment (Article 11) but assures the right to water in different other international conventions[25]. In addition, the D.L. N° 2.603 of 1979 introduces norms about water use and puts the nation's president in charge of the establishment of a legal system. Further, the Water Code (Código de Agua) regulates land water (surface and groundwater) by setting the substantive, procedural and administrative rules. The latter mostly sets the framework for the public institution DGA (cf. CNR and Universidad de Concepción 2016: 388). Besides those, the rules for private legislation, such as contracts, civil duties or private property are regulated by the Civil Code (Código Civil). In Chile, the legislation distinguishes between consumptive water (water that is not returned to the shed such as irrigation[26], mining[27], industrial and domestic use) and non-consumptive (water that will be returned such as water for hydroelectric generation) (Article 13, 14 Water Code). However, there is no prioritization of any use by the Water Code.

---

25 Convention on the Rights of the Child; Convention on the Elimination of All Forms of Discrimination against Women; Convention on the Rights of Persons with Disabilities; and, Convention No 161 of the International Labour Organisation of 1985.

26 Irrigation is regulated by Law N° 18.450 de Fomento al Riego, modified by Law 20.705 of 2013.

27 Water for mining is regulated by Law N° 18.248 de 1983, which sets up the Mining Code (Código de Minería) and the Law N° 18.097, Orgánica Constitucional sobre Concesiones Mineras.

The permissible forms of punishment in case of law violation are determined by the *Código Penal* (Penalty Code). Here, the most important articles concerning water are Art. 459 and Art. 461 which determine a financial punishment for illegal water withdrawal or water flow distraction of 11 to 20 UTM[28] (unidades tributaries mensuales) corresponding to a maximum of about 930.000 CLP (about 1260 € - calculations of UTM vary and depend on the inflation rate). Further, Art. 495 No.22 sets a penalty of one UTM for illegal water use or distraction of water flow that does not provoke a financial damage outreaching one UTM.

The rules of collective-choice level »concerning the assignment of monitoring responsibilities and specification of the magnitudes of sanctions« and the operational-choice level »governing implementation« (McGinnis and Ostrom 2014: 9) will not be separated as suggested by McGinnis and Ostrom but described in combination in the following. Due to the great overlapping of the two levels, a separation did not seem reasonable for the present study.

On a regional and local level water is managed by Water Users Organizations (WUOs). WUOs can be built by two or more persons who possess water rights of the same water body (river, groundwater, canal etc.). Their main management tasks are to distribute water, to administrate constructions and to resolve conflicts of trade water rights (MOP and DGA 2015: 120; Retamal et al. 2013: 13). In all different organizational forms, each member has the right of one vote for each water right he or she possesses (Article 222°, Water Code). The main organizational forms are:

- *Junta de Vigilancia (JV)* is a group of water users of the same basin. They usually consist of other WUOs, such as the Comunidades de Agua o Asociaciones de Canalistas described below. The Junta de Vigilancia aims at managing and distributing the water of one river basin. Hence, they obtain greater territory than the individual corresponding WUOs. As name Vigilancia (= monitoring) signifies, its objective is to provide mechanisms to mutually monitor law compliance. Further, decisions on constructing water infrastructure to improve resource extraction can be made and implemented by the Junta de Vigilancia after the authorization by the DGA (cf. Aldunce et al 2015). In

---

28 Unidad Tributaria Mensual (UTM) is the monthly fiscal unit used by the state to calculate taxes and fines. It is regularly updated by the inflation rate (Dectreto Ley Nr. 830, Art. 8). Here it has been calculated on the 26th of November 2018.

the Petorca Province, the governmental support of the formation of a JV has not been successful yet. Even though the bureaucratic obstacles for building a JV for the Petorca River have been tackled since 2010 (e.g. regularization of water rights), there is no JV working in the Province yet.

- *Asociaciones de Canalistas (AC)* are groups of water rights holders who share a hydraulic construction of common use with which they capture and channel water from its natural spring to their properties (MOP and DGA 2015: 120). In Petorca two ACs are registered at the Ministry of Public Constructions (MOP 2017). This number is misdirecting as many of the groups fitting this description are registered not as AC but as CASF.
- *Comunidad de Aguas superficiales (CASF)* is an organization of water right holders of the same natural surface source that capture, conduct and allocate water by a hydraulic construction of common use. In Petorca 88 groups of CASF are registered (MOP 2017).
- *Comunidad de Aguas subterráneas (CAST)* is a group of water rights holders of the same underground catchment sector or aquifer. In Petorca twelve CASTs exist according to the registrations at the government (ibid.).
- *Comité o Cooperativa de Agua Potable Rural (APR)* are groups of community members in charge of the provision of potable water in rural areas which have been abandoned by private water providers. Costs are regulated in the statutory of each committee or cooperative, and service quality is monitored by the Health Ministry (Servicios de Salud del Ambiente, Ministerio de Salud). The corresponding concessionary sanitary service companies, ESVAL in the case of Petorca, are legally obliged to technical and administrative counseling. Ten APRs are registered in the Cabildo District, twelve APR are registered in the La Ligua District, and 14 APR are registered in the Petorca District. They serve approximately 36.700 users (information obtained by MOP). However, there is an indefinite number of unregistered APR. For example, a report of the *Universidad de Playa Ancha* counted 15 additional informal APR in Petorca District in 2014 (UPLA 2014). Consequently, those informal APR neither benefit from technical nor administrative counseling, nor from other subsidies. Furthermore, experts state that APR suffer from scarce economic resources. As their income depends on charging rates, respectively on the number of supplied houses, more than half of the APR in Chile cannot cover their costs. Additionally, the educational level of their members and direc-

tors is rather low, which affects their management capacity. Technically, wastewater treatment seems to be the most outstanding problem as most of the APR do not have a sewage system (cf. Fuster and Donoso 2018).

[GS7 PROPERTY RIGHT SYSTEM]
To understand the property right system in Chile, it is crucial to know that water rights are separated from land rights (Art. 1, Código de Agua). Furthermore, even though Article 5 of the Water Code declares waters to be *»[...] bienes nacionales de uso público [...].«* *(»[...] national goods of public use [...]« translated by author)*, the constitution transfers the property to the private sector in Article 19 No.24

> »[…] Los derechos de los particulares sobre las aguas, reconocidos o constituidos en conformidad a la ley, otorgarán a sus titulares la propiedad sobre ellos; […]«. (»The rights of the private individuals over the waters, recognized or constituted by law, grant to the title holder the property over these; […]«, translated by the author)

The river basins in Petorca Province have repeatedly been declared to be an Àrea de Restricción (Restricted areas). Moreover, they were officially declared to be dried out since 1997 (River Petorca) and 2004 (River La Ligua) (cf. Art. 65, Water Code) (DGA 2018). This means that the National Water Directive (DGA) can only grant temporary water rights *(derechos provisionales)* in that area. Those temporary water rights can be turned into indefinite water rights after five years if they do not affect existing water rights (cf. Art. 66-68 Water Code). Three mayor problems concerning the water right system in Petorca were detected: Firstly, the DGA over-granted water rights. A study conducted by the University of Playa Ancha illustrates that the DGA was giving away over 40 % more water rights than available in the river basins (UPLA 2014: 32). An employer at the DGA of the Vth Region admitted the existence of those problems and insisted that there would be no more water rights handed over from the DGA to the private sector in Petorca (Interview 35).

Secondly, a great amount of water rights held by small- and medium-scale farmers are not registered correctly. The current situation of regularized property rights in Petorca is problematic. Experts estimate about 4000 not correctly registered water rights in that area. The main concern is that either small-scale farmers never registered their water rights after the introduction of the Water Code in 1981, or inherited water rights were not officially transferred to the successors. This situation of ambiguity does not only lead to family and small-scale community conflicts but also to bureaucratic obstacles when water right owners aim to join governmental programs (e.g. JV, CASF, CAST etc.), enter the water market or must justify their water use. Therefore, the state has commissioned two lawyers of the office of »*Programa de Gestión Hídrica*« (PGH) of the *Pontificia Universidad Católica de Valparaíso* (PUCV) in La Ligua to regularize about 700 water rights in Petorca in a project from 2015-2017. If the project is accepted by the communities and there is a demand for more regularization, the PGH will try to prolong the project (PUCV 2018; field observation).

The third problem consists of the process of change of water extraction points. The regulation granted water right owners the possibility to change their extraction point from the downstream area towards the upstream area. This process facilitates to obtain water under aggravated conditions such as dry spells. After 2014, however, the regulative system was modified to prevent such behavior as it raised water stress and conflicts in river basins (cf. CNR and Universidad de Concepción 2016: 384). The researchers of the CNR and Universidad de Chile (ibid.) demonstrated the distribution of extraction point changes of the ground water rights in the Petorca Province between 1988 to 2015 and show that during the megadrought (2010-2015) extraction changes increased drastically and then dropped after the regulative change in 2014.

[GS10 HISTORICAL CONTINUITY]
Taking the approach of critical institutionalism as a base, this subchapter displays path dependencies of the historical institutional development. However, it does not claim to explain the history of relationships or learning processes of the different actors, neither does it include informal traditions or culturally embedded local strategies. Such explanatory approaches are introduced in the following chapters of *actors and action orientation* and further discussed in Part 4. Rather than outlining the historical development of the actors' behavior, this subchapter concentrates on the historical continuity of the governance systems in terms of how the formal institutional framework has developed since its first Water Code in 1951 and which (Western) concepts have led to changes of the institutional construct in which the actors move.

Carl J. Bauer, expert on Chilean water rights, describes the historical continuity by the metaphor of a pendulum that started with rather equalized powers between state and privates in 1951. Then, it swung towards nationalization in the late 60s and then towards intense privatizations at the beginning of the 80s where it has remained until today (Bauer 2015:

73). In 1951, the Law N° 9.909 was introduced, and today it is considered to be the first Chilean Water Code. This law allows the government to grant provisional water rights that would become protected property rights after being used. As Carl. J Bauer states:

»[...] 1951 Water Code combined private water rights with strong government regulation, in a way that resembled the Western United States in the early 20th century.« (Bauer 2015: 151). First, water management was administrated by the Ministry for Public Work (Miniserio de Obras Públicas); later in 1969 the National Water Directive (DGA) was introduced. Unlike today, the regulative system restricted the power of private entities in several ways: inter alia it prioritized water for domestic use and allowed the DGA to withdraw unused water rights after five years. Further, water rights applicants had to specify their water use; definitive water rights were only given away after norm compliance assurance, and water rights were undividable from land rights (Bauer 2015: 85). Supported by the US president John F. Kennedy, the Chilean center-left government of President Eduardo Frei M. introduced an emblematic reformist policy. Concerning water management, the most outstanding reform was introduced in 1967: An Agrarian Reform Law that strengthened the governmental power by land expropriation and redistribution of land and water rights to its occupants (ibid.: 151). This process has led to administrative confusion in terms of unclarified property rights up to today (ibid.: 88). Moreover, the DGA took over responsibilities from the law courts reducing their power. From 1970-1973, Chile was governed by the socialist president Salvador Allende who aimed at profound nationalization of resources and governmental power over the economic market.

After the military coup in 1973 several reforms were carried out that transferred the management system towards a water market and ended in the introduction of a new Water Code in 1981. In order to understand this water code, according to Bauer (2015), it is crucial to understand its origin and its dependence on the Chilean constitution. Both, the water code of 1981 and the Chilean constitution of 1980, follow paradigms of the free market. The constitution was introduced during a military dictatorship that did not allow any opposition or participatory process. Until today, a full agreement of the congress is needed to change that constitution. Hence, any reform of the water code will hardly alter the paradigms carved in stone in the constitution. To assure economic liberty in the water code of 81, priority of water for domestic use was abolished, and water rights were handed out for free by the DGA without specification of use. The water code of 81 does not include environmental protection or water quality assurance (ibid.: 73-105). However, it should be mentioned that the recent history of Chilean water rights has also been told from different points of view. The Chilean lawyer and expert in water rights Alejandro Vergara, for example, recognizes the profound connection between constitution and water code. However, he concludes that the constitutions in addition to protecting the property of water rights and classifying their origins into constituted (by administrative concession) or recognized (by customary practices), assures water to be a public good. He equates public good with common good and takes that as an argument to oppose the privatized character of water that has been attested by others (Vergara Blanco 2014: 88-89).

In 1988 the military dictatorship was ended by plebiscite. However, the preconditions for this plebiscite were inter alia that the constitution of 1980 remained. Although the following central-left governments tried to introduce several reforms, the opposition used their possibilities to block them successfully (Bauer 2015: 260-261). I argue that privatization of water management went even further, as the 16 public water supply companies were privatized and given over to transnational companies during democracy especially during the governments of Eduardo Frei (1994-1999) and Ricardo Lagos (1999-2005). In 2004 almost all water rights for human consumption were in the hands of the private sector. In 2005 a water reform towards more democracy and sustainability was introduced as well as a fee for unused water rights and enhanced environmental rules (Larraín 2010: 8; Clarvis and Allan 2014: 83). The latter only apply on water rights that have not been handed out yet. Consequently, criticism and claims for further reforms remain. Valdés-Pineda et al. (2014) suggest improvements, mainly addressing a power-shift towards the state and participative strategies, whereas Retamal et al. (2013) and Larraín (2010) claim that the entire institutional system is insufficient. According to them, participation is looked upon as part of democratization, but still treated more as an obstacle which has to be faced than as a tool for development (Retamal et al. 2013, Larraín 2010) Moreover, Retamal et al. (2013: 12) conclude that the institutional change since 2005 has not improved public participation at all; instead, they describe that the social demand for participation is seen as a threat to the state. Clarvis and Allan (2014: 88) state that the recent strategies in Chile focus mainly on technical solutions while paying too little attention on user participation. One governmental strategy to implement this issue is the development of public private partnerships on watersheds. Since governmental opportunities to install such cooperation are highly restricted due to the water code, these partnerships have been criticized for being very weak (Retamal et al. 2013: 14).

Starting the political conversation in 2010, in 2015 the congress members (diputados) voted in favor of a new water code reform covering, inter alia, the following aspects: prioritization of water for human use, protection of glaciers, end of infinitive property rights, implementation of »ecological basins«. While this step has been celebrated especially by Chilean NGOs associated with the political left (Chile Sustentable, Observatorio Popular, Ecosistemas), the reform must still pass through the chamber of senators and other political negotiations (field observation).

7.5 Actors and action orientation

This study focuses on the actors' behavior and their relationship to the institutional system in a water conflict. As an analytical framework I used AI by Mayntz and Scharpf (1995). In order to consider the interdependencies of the focal situation together with the surrounding context, the AI was inserted into a greater SESF based on the model developed by Ostrom (2007, 2009). The following qualitative empirical data were collected by interviews and field observations and form the core source for arguments in the discussion of the actors' behavior. Therefore, chapter 7.5 consists of a detailed, in-depth data analysis and is the largest chapter of Part 3.

[A1 RELEVANT ACTORS]
Firstly, the relevant actors were identified in order to understand the actors' behavior and outline their primary characteristics concerning water governance. The list of relevant actors is similar to the list of the rule making organizations showed in chapter 7.3. However, while the rule making organizations show a general overview of actors directly or indirectly influencing the rules of the focal situation, the following characterization specifies the relevant actors who influence, respectively are involved in the water conflict in Petorca. According to Ostrom (2007) actors can be divided into two categories: direct users of natural resource and other actors. In Petorca direct users are domestic water users (households), agricultural water users and industrial water users, while other actors are mainly water providers and the various governmental bodies. The group of domestic water users divides into urban water users provided for by the private company ESVAL and rural water uses provided by APRs. Agricultural water users are the user group with the highest water demand. Ap-

proximately two thirds of all farmers are male and most of the farmers have a low level of education (basic education). Their age ranges mostly between 60 and 70 years (CNR and Universidad de Concepción 2016). While more than two thirds of the farmers in Petorca are small-scale, large-scale farmers own most of the land (ibid.). Further, large-scale farmers are organized in an association here called LSAA[29]. Other industrial users are small- and moderate mining companies located mainly in the districts of Cabildo and Petorca. Some actors (mainly domestic users and small-scale farmers) are organized in a movement called MODATIMA (Movement in Defense of Access to Water, Land and Environmental Protection). They cooperate with different NGOs on a national and international level. Further local players are academic institutions such as the *Programa de Gestión Hídirica* (PGH) that carries out projects related to water and agriculture in the province. Among the group of governmental bodies and authorities, the municipalities, and the Institute for Farming Development (INDAP) were identified as relevant actors on a local level. On a regional and national level, the Ministry of Agriculture (MINAGRI) and the Ministry of Public Work (MOP) were recognized. In specific, the following departments play significant roles: National Irrigation Commission (CNR), National Water Directive (DGA) and Directive of Hydraulic Works (DOH). Additionally, the National Commission for Hydrologic Issues (Comisión Nacional de Asuntos Hídricos), which was founded in the congress, was important to this study.

The relevant actors and actor groups identified were clustered into different social units. The social unit is important to understand the actor's role and its *»role-specific norms and expectations, which will also define the social unit that is to be served by role-specific actions.«* (Scharpf 1997: 61) Thus, the actor's behavior corresponds to the normative expectations connected to the social unit rather than to individual preferences (ibid.). I categorized the actors into cooperative actors that *»[...] are acting in a top down structure, the beneficiaries of their action are highly independent of their own benefits and action is carried out by staff members [...]«* (Scharpf 1997: 52-58), collective actors that *»[...] are mostly informal organized and have the same action orientation, they are guided by members preferences [...]«* (ibid.) and aggregated actors who *»[...] are those which do not fit in the description of cooperative nor collective but are related to each other e.g. in networks or family structures.«* (ibid.) as

---

29   Name is fictitious.

explained in chapter 4.2.1. This categorization helps to evaluate the potential general validity of the generated information and to explain disagreements within groups. Table 9 shows actors and actor groups that were recognized as outstandingly relevant to the case and gives and an overview of the actor groups that were analyzed in the following subchapters.

Table 9: Selected actor groups for interview analysis. (Own research)

Group	Subgroup	Type
Society	Citizens	Aggregated
	Activists	Collective
	Local leaders	Cooperative / Collective
Governmental authorities	Local	Cooperative
	Regional	Cooperative
	National	Cooperative
Economy	Large-scale farmers	Aggregated
	Large-scale farmers association	Collective
	Water provider	Cooperative
Other	NGO	Aggregated
	Church	Cooperative
	Academic institutions	Aggregated
	Further experts	Aggregated

[A3 ACTION ORIENTATION]

The actors and actor groups are characterized by different perceptions, identities, motivations that are influenced by their social-economic status and individual experience. As described in chapter 4.1, this chapter relies on close readings of in-person, semi-structured interviews with individuals of different actor groups as well as field observations in order to show the actors' orientation. The interview guide was based on the sub-tiers developed for the category [actors] of the SESF by Ostrom (2009) and del Mar Delgado-Serrano and Ramos (2015). Furthermore, sub-tiers were extended, respectively substituted, by indicators for action orientation as described by Scharpf (1997).

Taking this conceptual framework as a base, I developed different actor group mind maps that helped to explain the actors' behavior and to structure each actor, respectively actor group, analysis into (1) cognitive aspects referring to the actor's perception of the situation causality, (2) motivational aspects referring to norms, interests and identity and (3) orientation of interaction. To define the orientation of interaction of the interview partners, I concentrated on statements about experienced interaction and on statements with a judgmental character that revealed their feelings towards other actors or actor groups. Scharpf describes different interaction orientation with a function that calculates the payoff:

»$U_x = aX + bY$ where U_x is the total utility that is subjectively experienced by ego; X and Y are the »objective« payoffs received by ego and alter, respectively; and a and b are parameters varying between -1 and +1.« (Scharpf 1997: 95)

In the following subchapters, I used this function to describe the interaction orientation with the actor groups that were mentioned by the interview partner. I named the actor groups *civil society,* divided into *citizens, activists,* and *local leaders; governmental authorities,* divided into *local, regional* and *national authorities,* and *economy* divided into *large-scale agriculture* and *water provider.* Moreover, there is a group called *others* including various experts from organizations such as academics, church, NGO or the private sector.

7.5.1 Actor group: civil society

The interview partners categorized as *civil society* are citizens whose center of life, such as work and/or family, is in in the Petorca Province and who are not representatives of governmental bodies or large-scale private industry. Most of the interviewed people in this category are small-scale farmers. Some may argue that individuals of *civil society* belong the actor group *economy* since small-scale farmers form a large part of the local economy in terms of employment numbers. However, I argue that Petorca' s society is shaped by small- and medium agriculture. Thus, I preferred to show their action orientation through the different subgroups of society rather than considering them as a subgroup of economy. The subgroups of civil society were formed according to the role in society of the corresponding interview partner. I distinguished between *local leaders, activists* and *citizens.* The latter are persons without a formal or informal leading role or activist identity. Further, I added field observations made in different WUOs meetings to this analysis.

7.5.1.1 Local leaders

The actor group of *local leaders* are society members that fulfill leading positions in local WUOs. The four interviews in this category were held with four presidents and one secretary of either Canal Associations (AC) or Potable Water Associations (APR). I categorized them as collective actors because their actions are guided by their members' preferences, and they do not act independently for their own benefit (cf. Scharpf 1997: 57). The following Table 10 provides an overview of basic interview details:

Table 10: Local leaders. (Own research)

No.	Name (fictitious)	Sex	Position	Age	District	Profession
1	Alejandra	F	AC President	50-64	Petorca	Farmer
	Rosa	F	AC Secretary	50-64	Petorca	Farmer
2	Luis	M	AC President	80-95	La Ligua	Farmer
3	Ana	F	APR President	35-49	Petorca	Unemployd
4	Sergio	M	APR President	65-79	Cabildo	Farmer

OVERVIEW

Concerning the motivational and cognitive aspects, all interview partners are personally affected by the water scarcity. They are known as leading persons and approached by their community in search for help on different occasions. They are proud to support their community and show a heroic attitude of dedication and sacrifice. All participants consider overexploitation of water by large-scale agricultural industries as the main reason for the water scarcity while they recognize climate change as a reinforcing factor. Further, they blame the water regulation for being too weak and complain about unfair laws that favor the industry and believe that this situation roots in the downfalls of water privatization. Likewise, they have a negative perception of governmental programs addressing scarcity problems in Petorca. Furthermore, they express great concern about the lack of interest in cooperation by their community. According to the interview partners, cooperation and organization of the community is the most suitable way to finding a solution of the water conflict. Other actor groups mentioned by the *local leaders* are *large-scale agriculture, governmental bodies & authorities, activists* and *civil society*. The analysis identified selfish-rational behavior as the prominent interaction orientation. Additionally, competitive behavior was observed towards *large-scale agriculture*. Only towards the *civil society*, the *local leaders* show a more cooperative behavior. In specific, solidary behavior was detected towards *WUOs*. The following part shows a detailed analysis of the motivational and cognitive aspects and the orientation of interaction.

MOTIVATIONAL ASPECTS

All *local leaders* are personally affected by the water scarcity. They or their families have been damaged economically because of crop losses and the costs of buying irrigation water from neighbors or because of technological changes (Interviews 1, 2, 4). One interviewee was provided with potable water from trucks that are financed by the municipalities (Interview 3).

Furthermore, all interviewed people are known by their community as *local leaders*. This means that community members recognize them as outstanding persons who can be approached in case of problems - even if the issue is not within the WUO's remit. For example, Sergio says that in case of certain problems, they contact him, and he then contacts the police (Interview 4), and Ana claims to have been on television demanding the streets to be paved (Interview 3). During the interview, Rosa repeatedly recognizes Alejandra as a community leader, e.g.:

> »Gracias a ella se ha logrado y bueno el grupo que la seguimos. [...] Si ella no hubiera hecho nada, el pueblo no estaría con agua de nuevo.« (Interview 1)

> (»Thanks to her, it has been successful and well, the group of us who is following her. [...] If she wouldn't have done anything, the village wouldn't have water again.« Translated by author)

At the same time, the *local leaders* identify with this role by proudly showing their sacrifice. They see themselves as community servants or volunteers and often believe to be one of the few persons who really care for the people as this statement by Sergio shows: »*Yo creo que soy uno de los pocos que defiende a la gente acá*« *(Interview 4) (»I think I am one of the few who defends the people here.« Translated by author)*

While complaining about this situation and demanding more cooperation within the communities, most of them repeatedly underline their willingness to help, their restlessness and their networks. They stress that they would like to quit their positions in the WUO, but that there is no one who would like to take over their role (cf. Interviews 1, 2, 4). By this, they contrast their actions to the behavior of their community members and give the impression of being proud of their outstanding, in some cases even heroic, role.

COGNITIVE ASPECTS

To understand the conflict perception of the actors, it is crucial to explore their causality patterns. In addition to their personal damage, they point out impacts of the water scarcity on the community. They recognize the selling of land and water rights, the employment loss in the agricultural sector and the economic damage in the province as consequences of the water scarcity and as driving factors that cause migration into other regions. Climate change is only rarely mentioned by Alejandra as one of the principal reasons for water scarcity (Interview 1). Ana mentions climate change impacts only after being asked about them (Interview 3). Luis and

Sergio do not refer to climate change at all but mention the lack of rain over the last years as a reason for the scarcity (Interviews 2, 4).

Most of the interview partners argue that overexploitation by large-scale agricultural industries, privatized water system and incompetent or corrupt governmental bodies have caused the water scarcity and conflict. They repeatedly criticize the overexploitation of water resources by large-scale agricultural companies (cf. Interviews 1- 4). Ana, who is neighbor to a large-scale agriculture entrepreneur, answers the question about the scarcity's reasons as follows:

»Las razones están porque en los que tienen poder que son los que tienen más, se llevan el agua. Y se dejan a nosotros sin agua. Porque […] hacen unas construcciones de aguas, que hacen unos pozos profundos y eso hace que el agua se consuma y que ellos hacen pozos y de los pozos el agua se disminuya y ellos se quedan con el agua.« (Interview 3)

(The reasons are there because the ones who have the power, which are the ones who have more [money], take the water away. And they leave us without water. Because they make water constructions, deep wells and that causes that the water is consumed, and they build wells and the water decreases and they end up with the water.« Translated by author)

This statement indicates that the water access depends on the socio-economic status and is, consequently, unequal. Sergio confirms this by saying: »*Aquí dicen que el agua es de todos, pero él que tiene plata la saca.*« (*interview 4). (Here they say that water belongs to everyone, but only the one who has money gets it.*« Translated by author) Most of the *local leaders* see this situation connected to the water rights privatization which they strongly criticize (Interviews 1, 2, 4). They wish that water would be a common good owned and shared by the community (Interviews 2, 3, 4). Further, they express their concerns about the lack of sanctioning water misuse (Interview 4), about laws that would favor large-scale entrepreneurs (Interviews 1, 3) and about large-scale agricultural companies that influence governmental bodies (Interview 4). These concerns are part of the general perception of an unfair system. The interviewees make accusations reaching from unfair distribution of funds to corruption (cf. interview 1, 3), witness the following example:

»De 1 a 10 estamos en el 9 con la corrupción. […] Del gobierno, de la municipalidad de gobernación, todos, incluyendo toda el área nuestra. Todo, el país, los municipios que tenemos.« (interview 1).

(»From 1 to 10 we are at 9 concerning corruption. [...] Of the government, the municipality, the gobernación [regional government], all of them, including all of our area. Everything, the nation, the municipalities that we have.« Translated by author)

This negative perception of the Chilean governmental system is fueled by their experiences with governmental programs. The actors report to have received either no or too little help by the government in the past (Interviews 1, 3, 4). Also, long-term solutions offered by the government are poorly accepted. For example, the desalinization plant proposed by the mayor of La Ligua is regarded as too expensive (Interview 1) and ineffective and the Junta de Vigilancia is considered to be inconvenient for small-scale farmers (Interview 4). However, the interview partners mainly agree that the government should act more to alleviate the water scarcity conflict (Interview 3). Alejandra specifies that by demanding a change of the system, respectively the law. She criticizes that the juridical system has not changed during democracy, and she has little hope for the future because she believes that powerful large-scale entrepreneurs will oppose against such a change (Interview 1).

Another issue that worries the interview partners is the lack of interest in cooperation of their community members. They state that people do not show up for meetings (Interview 4), that community members are not willing to take on responsibility (Interview 1), and that they prefer individual solutions to the water scarcity such as wells instead of mutually maintaining the canal (Interview 2) as they have lost hope of being able to mutually solve the water scarcity problem (Interview 1).According to the *local leaders*, especially lack of interpersonal trust restricts cooperation (cf. Interviews 1, 3).Alejandra claims that everyone suspects hidden interests and that people do not trust in the goodwill of others. Ana points out another problem that restricts cooperation and reports that organized persons are prone to be rejected by governmental authorities because they are considered as conflictive (Interview 3).

In other words, the interviewees assert a lack of trust inherent to society and from society towards governmental bodies. In response to that, all interview partners insist that people should cooperate more. They see community organization as an important way to cope with the water scarcity (cf. Interviews 1-4). As leading persons in different WUOs, their success depends on cooperation and compliance within the community. Hence, this importance on community organizations correlates with the aims of their social unit. Further suggestions for a conflict solution mentioned by the interview partners are education (Interview 1), raise of awareness by outsiders (Interview 3) and rain (Interview 4).

INTERACTION ORIENTATION

First, the interview partners describe their interaction orientation towards *large-scale agriculture* that is characterized by negative connotations. For example, Rosa (Interview 1) calls large-scale agricultural entrepreneurs »fachos« (fascists). Ana (Interview 3) believes that leaving poor people without water does not concern them. This orientation is rooted in their causality patterns as they hold water overexploitation by large-scale farmers responsible for water scarcity. Cooperation with representatives of the actor group *large-scale agriculture* have mostly failed, as the following examples illustrate: Sergio and Ana are neighbors of large-scale agricultural companies. Sergio, despite claiming to have a rather good relationship with his neighbor, reports that agreements on water use with his neighbor only happened after going to court and that currently he is suing him for illegal water extraction again (Interview 4). Ana (Interview 3) also reports about failed cooperation with her large-scale farming neighbor. According to her, even in the worst times of drought, he was not willing to share water with her WUO. Currently, she has no contact with him.

Summing up, the interview partners have a rather negative - in some cases even hostile - image of the representatives of *large-scale agriculture*. Therefore, I conclude that the interaction orientation of *local leaders* towards *large-scale agriculture* can be described as mainly competitive. The *local leaders'* negative image of large-scale agricultural entrepreneurs is mostly interlinked with their perception of an unfair and corrupt government. These cognitive aspects together with their own interests (dependence on irrigation and drinking water), in addition to their feeling of being responsible and wanting to help the community as *local leaders*, create a competitive situation about the scarce resource. Nevertheless, their solution strategies show that they do not behave in a hostile way towards their alter. They are not primarily interested in their loss but try to find benefits for themselves or for their community. Hence, rather than a hostile interaction orientation ($U_x = -Y$), a competitive interaction orientation dominates that evaluates a gain of the ego equally as a loss of alter ($U_x = X - Y$). In the case of Luis, individualistic/ selfish-rational behavior focusing on the egos benefit ($U_x = X$) was found. He shares his negative view on water privatization as well as the experience of unequal water access with the other interview partners. Yet, he does not refer to any competitive relationship with large-scale agricultural companies.

The second group mentioned by the interview partners is *governmental bodies & authorities*. All interview partners cooperate with governmental bodies such as DOH, DGA, INDAP, MOP and municipalities to obtain subsidy for technical upgrades (Interviews 1-4). This cooperation is part of their role as leading persons in WUOs and therefore of personal as well as of community interest. As explained in the cognitive aspects above, they perceive the regulative system as unfair, especially because of the perceived closeness between governmental bodies and big economic players (Interviews 1, 3, 4). Therefore, and because of failed cooperation in the past, they have no or little trust in governmental institutions (cf. Interviews 1, 3). Ana, for example, relates:

>»Ahora ya no confío en nada porque ahora ya hemos peleado tanto que ya estamos un poco ya desilusionado porque [...] de repente hacen cosas por prender la política no más. Dicen que si y después no, lo dejan ahí.« (Interview 3)

>(»I don´t trust anything anymore, because we have fought so much now; we are disappointed, because [...] sometimes they make up things only to light up politics. They say yes and later no, they just leave it as it is.« Translated by author)

Rosa reports a case in which the municipality first wanted to help her, but then the funding was inexplicably lost. According to her, such experiences do not only nurse the mistrust of the community members who therefore avoid participating in meetings (Interview 1). Ana reports that her dependency on the help of governmental bodies makes her feel powerless and humiliated:

>»Entonces vemos tanta agua y al final nosotros nos sentimos un poco impotente, un poco humillándonos de repente, tenemos que como mendigar para ver que nos pongan el agua.« (Interview 3)

>(»So, we are seeing so much water and, in the end, we feel a little powerless and a little humiliated sometimes, we have to beg them to provide us with water.« translated by author).

While Ana remains skeptical towards governmental bodies, Sergio's interpretation of the authorities' behavior goes further. He thinks that the governor does not like him personally and works actively against him together with his neighbor, for example, by not inviting him to meetings (Interview 4). As these statements indicate, the interview partners mostly have a negative image of *governmental bodies & authorities*. Nevertheless, they feel dependent on them. Thus, they are not interested in their loss. A hostile or a cooperative interaction orientation can be excluded. Their interaction orientation towards *governmental bodies & authorities* can rather be classified as selfish-rational and individualistic because they cooperate only for their own or their communities' benefit ($U_x = X$).

The next actor group discussed by the interview partners is *activists*. Sergio considers himself an activist because he works together with MODATIMA. He says that the group fights for the same interests as he does, and therefore he believes to be a link between them, the community and governmental bodies (Interview 4). Thus, his interaction orientation can be described as cooperative or solidary, as defined by Scharpf (1997: 85) as $U_x = X + Y$. In this case as, »*A gain to alter or a gain to ego will be equally valued.*« (ibid.) The further interview partners of this group have different views on the activists. Luis (Interview 2) relates, that »*[...] nos han hablado en reuniones de que no sea privada*« (*[...] they have talked to us in meetings about that [water] should not be privatized*). Apart from agreeing with this statement, he does not mention any further relationship. Alejandra, in contrast, has participated in various events organized by MODATIMA, and does not trust them or any social movements in general. She thinks that especially MODATIMA is a closed group which is not easy to work with (Interview 1). She concludes that »*[...] ni nos ayudan ni nos molestan*« (*[...] they neither help us, nor they disturb us*) (Interview 1). In that sense, Alejandra has a selfish-rational, individualistic interaction orientation ($U_x = X$), as she does not care about the benefit or loss of the alter but concentrates on her own win.

The personal benefit of the *local leaders* depends heavily on the benefit of their community. As WUO leaders, they are expected to work for the welfare of their community members. The interview partners report that they successfully cooperated with other WUOs by either offering or receiving help (Interviews 1-3). With regard to cooperation within their own group, the interview partners stress that there is a lack of cooperation, lack of trust and hope as well as preference of individual solutions restricts the communities' organization. Therefore, I conclude that the interaction orientation of *local leaders* towards *civil society* and especially WUOs is cooperative and solidary ($U_x = X + Y$). Whereas, within the broader group of *civil society* and WUOs the interviewed persons show a rather selfish-rational, individualistic behavior ($U_x = X$).

7.5.1.2 Activists

This subchapter focuses on the actor group *activists*. All analyzed interviews partners were, respectively are, part of the activist group MODATIMA. In 2000, the group emerged from an initiative of residents in Petorca Province that consisted of mostly small-scale farmers who joined to fight water privatization and to reclaim water as a common good. With regard to their social unit, they were categorized as collective. Although the group registered as a cultural organization in 2009, they claim to have no organization structure and describe themselves rather as an open movement (cf. Interview 6). Their members are led by what they believe is best for the group and for the residents of Petorca Province. The Table 11 below lists some basic information about the interview partners' sample structure:

Table 11: Activists. (Own research)

No.	Names (fictitious)	Sex	Position	Age	District	Profession
5.	Gabriel	M	Activist	50-64	Cabildo	Farmer
6.	Juan	M	Activist	50-64	Petorca	Construction worker
7.	Mateo	M	Ex Activist	50-64	Petorca	Unemployed
8.	Pablo	M	Activist	50-64	La Ligua	Agronomist

OVERVIEW

On the whole, all activists are personally affected by the water scarcity - either with regard to drinking or irrigation water. The motivation for their activism rests on their concerns about a sustainable future and their strong connection with the Petorca community, in specific with small-scale farmers. According to the interview partners, the downside of their identity as activists is intimidations by politicians and industry as well as social exclusion. The cause of the water scarcity is seen equally in climate change and in overexploitation by the large-scale agricultural industry and as caused by water privatization and weak water regulation. In that context, the activists emphasize a strong connection between the government and the industry. Like the *local leaders*, the *activists* are concerned about the lack of cooperation in society. In addition to the mistrust and the feeling of powerlessness, the interview partners mention fear as a discouraging factor. The orientations of interaction range from hostile, competitive, selfish-rational to cooperative. Hostile orientation dominates towards *large-scale agriculture* and *private water providers*, whereas towards *governmental bodies and authorities* hostile and selfish-rational orientation were observed. Furthermore, the interviewed persons show cooperative orientation towards *civil society* and other *activists'* groups. The following part provides a detailed analysis of the interconnections of the activists' motivational and cognitive aspects as well as their orientation of interaction.

MOTIVATIONAL ASPECTS

The movement MODATIMA is spread all over Chile. However, all interviewed activists are residents of Petorca Province and are personally affected by the water scarcity. Not only do Gabriel and Mateo share the lack of drinking water and their concerns about water quality, they also suffer economically as they are working in the agricultural business. Mateo, for example, built a cooperation of small-scale farmers that exported avocados. But this group does not longer exist because of the water scarcity (Interviews 5, 7).

»*[...] la lucha por el agua es la defensa de la vida [...] (the fight for water is the protection of life)*« (Interview 8). Following this topic, the interview partners stress the importance of water and its fair and sustainable management for their own future as well as for future generations (Interviews 5, 6, 8). Therefore, they identify with their role of being a social leader and/or water activist who fights to reclaim water (interview 6 and are recognized as water activists by their community (Interviews 5, 6). Despite positive connotations towards being an activist (for example Juan feels empowered by the group), the interview partners also comment downsides of this identity as it goes along with a reputation of being a conflictive person and has led to certain forms of intimidation and social exclusion (Interviews 5, 6, 8). According to Mateo, a founder of MODATIMA, the group has developed a violent image for what he is no longer an active member (Interview 7). Moreover, the activists do not belong to any political party. Instead, most of them claim to be politically independent (Interviews 5, 6). Furthermore, they feel strongly connected to the Petorca Province and to farming (Interviews 6-8).

COGNITIVE ASPECTS

While almost all interview partners recognize climate change as a reason for the water scarcity, all of them hold also large-scale agricultural companies responsible. According to them, large-scale agricultural companies overexploit the water resources (Interviews 5-8). They state that those entrepreneurs harm the most vulnerable parts of society: poor residents and small-scale farmers who cannot effort to buy new technologies or build deeper wells (Interview 6, 7). In addition, they accuse large-scale agricultural business of illegal water extractions (Interviews 5, 7, 8). The interview partners argue that this water robbery is ascribed to the power of companies over governmental authorities, alongside with (legally) weak governmental bodies, specifically the DGA (Interview 5-7). The interviewees state that laws favor financially rich people, that regulations are lacking, and that the level of corruption is high (Interviews 5, 6). They believe that the strong connection between government and industry is accountable for those weak regulations which leads to an unequal access and distribution of the resource (Interviews 5, 7, 8). Furthermore, they claim that the governmental bodies are led by incompetent politicians (Interview 5, 7), with a predominant win-maximizing ideology (interview 8): *»[...] hoy en día no manda el pueblo, mandan los empresarios [...] (Interview 6) (»These days not the people but the businessmen are in charge«* translated by author*)*.

All activists the water code implemented in 1981 as a starting point for the current mismanagement. They comment that by privatizing water, the resource was handed over to the industry and therefore, equal access to it and its distribution were hindered, and its capitalistic uses were given preference instead of its use for human wellbeing (Interviews 5, 7, 8). Consequently, the activists seek for a new regulation through which water will be treated as a common good (Interviews 6, 8). Further, most of the interviewees are not satisfied with the governmental programs on water management. To give some concrete examples, Gabriel criticizes the desalinization plant as a tool for covering water robbery by trying to provide more resources instead of dealing with the scarcity causes. Moreover, he says, that the conflict resolution measure offered by the government called »mesa técnica« (or water round table) was doomed to failure from the beginning. He criticizes that this discussion round was led by the large-scale farmers association LSAA and that none of the activists or people who denunciated water robbery were invited to these meetings (Interview 5). Mateo complains about the »Ley de Mono«, a law introduced in 2005 to give farmers the possibility to legalize their wells, which was then taken advantage of by large-scale agricultural companies.[30] In contrast to the other activists, Mateo is in favor of the governmental program Junta de Vigilancia. He has supported the project for several years. According to him, its implementation failed because of the mismanagement of the main organizers of the *Progama de Gesión Hídrica* (PGH[31]). Moreover, he is the only one who highlights technical solutions for the water scarcity such as renewable energies (Interview 7).

---

30 For more information on »Ley de Mono« and »Mesa Técnica« please read chapter 8.2
31 For more information on PGH please read chapter 8.2.

Overall, the activists seek for a profound change towards a sustainable and socially just water management, for example, in terms of a constitutional change (Interview 7) or other mechanisms that would assure norm compliance (Interview 5, 7). As described in the motivational aspects, the interview partners feel deeply connected to the citizens of Petorca Province – especially to small-scale farmers. Therefore, they all agree on the importance of people organizing or working together in order to achieve change. At the same time, the interviewees complain that the people in Petorca Province are not interested in such cooperation. The interview partners believe that this lack of interest relates to the lack of information and education of society (Interview 5) as well as to a lack of environmental awareness that hinders citizens to understand their situation and its context (Interview 7). In addition, Juan reports that people feel powerless against large-scale agriculture industry:

> »La gente no está ni ahí, dicen »que le vamos a ganar a los empresarios« »que esa es una tontera«, »que las cosas ya están así y hay que dejarlas así«, eso es lo que dicen, »andan puro molestando no más hay que dejar las cosas así.« (interview 7)

> »The people don´t care, they say 'how are we going to win against the businessmen' 'this is nonsense', 'That's how things already are and you have to leave it like that.', that's what they say, 'they [activists] are only disturbing, you have to leave things like it is.'« Translated by author)

Additionally, the interviewees report that the people are disappointed of achieving any positive outcome from organizational meetings. Consequently, they lost motivation to participate (Interview 5, 7). Further, the interview partners say that people are afraid of organizing against large-scale agricultural companies and the current water management system, because, firstly, people work for those companies and do not want to lose their job, and, secondly, the fear is nursed by different kinds of intimidations and social punishments that activists are confronted with. To illustrate this point, the activists report about different attempts of intimidation. Gabriel says that he has been excluded from official meetings and was accused of illegal water distraction. He thinks that this is a strategy to make his accusations about others sound implausible (Interview 5). Juan points out that he gets called »communist« by his opponents and by other community members. He sees this as a strategy to harm his reputation and to silence his accusations (Interview 6). Pablo (Interview 8) refers during the interview to scare tactics and juridical persecution that he and other activists are subject to.[32]

ORIENTATION OF INTERACTION

The groups of interaction mentioned by activists are large-scale agriculture, private water provider, governmental bodies and authorities, civil society and activists.

Since the activists blame *large-scale agriculture* for the water scarcity, a negative and even hostile orientation can be observed ($U_x = -Y$). For example, they call the owners of large-scale agricultural companies shameless (Interview 7) and believe that they would have no qualms to leave poor people without water (Interview 6). Juan and Pablo shame the agricultural companies who were fined for illegal water withdrawal and who according to them keep on stealing water by repeatedly listing the names of (Interviews 6, 8). The interviewees have reported water robbery to the DGA and to the local court several times (Interviews 5, 8). Mateo (Interview 7) says that large-scale agricultural entrepreneurs repeatedly have lied about helping the community. Because of such experience as well as intimidation attempts, the activists claim to be unwilling to communicate with representatives of large-scale agricultural companies. According to Juan (Interview 6), large-scale farmers would rather like to see him dead than to communicate.

With regard to public water providers, Gabriel is the only one who explicitly mentions his relationship with ESVAL, the private water provider of Petorca Province. His orientation can be described as hostile ($U_x = -Y$). He expresses his anger about the company repeatedly. According to him, the company prioritizes water provision for tourist areas. Furthermore, he reports that in a fight between him, other farmers and ESVAL, he got trapped by the company and later sued (Interview 5).

The orientation of interaction towards *governmental bodies and authorities* varies between hostile ($U_x = -Y$) and selfish-rational ($U_x = X$). While one interview partner shows cooperative interaction orientation ($U_x = X + Y$), in general, the interview partners have a negative image of politicians

---

32   Such intimidation tactics became internationally famous when Amnesty International started a call for the activists protection after the activists publicly stated to receive death threats by phone and physical attacks (Amnestía Internacional 2018). In 2004, a member of MODATIMA was arrested and charged for publicly accusing the former minister of interior Edmundo Perez Yoma of stealing water (EL CIUDADANO 2014).

and governmental authorities. They believe that the political system is corrupt and unfair and, accordingly, they perceive the government's representatives negatively. For example, Gabriel (Interview 5) repeatedly gives examples of recent corruption scandals and conflicts of interests. He and Juan (Interview 6) claim that governmental authorities do not respect citizens because they have refused to listen to them at various occasions. The following example of Gabriel's experience in congress meetings illustrates his negative image:

> »[...] yo diría que nos ha pasado sus seis o siete veces, en el cual vamos con la mejor voluntad del mundo a exponer lo que ha pasado en nuestra provincia para tratar de ayudar a otras provincias, pero también a tratar de ayudarlos a ellos a que tomen algunas decisiones y ellos se paran con un desparpajo y se van o agarran sus teléfonos.« (interview 5)

> (»[...] I would say it happened about six or seven times that we presented with the best will of the world what is happening in our province in order to help other provinces, but also to help them to take some decisions and they impudently stand up and leave or grab their phones.« translated by author)

Because of such experiences, Juan (Interview 6) concludes that the authorities do not care about problems of the society. Pablo (Interview 8) states to have »the worst opinion« about the government: *»creo que las autoridades son una vergüenza.« (Interview 8) I think that the authorities are an embarrassment.« Translated by author)* According to him and Gabriel, the authorities are accomplices of the water robbery as they neither oppose the (in their view) abusive system, nor comply with the few existing rules in water management, nor monitor water management sufficiently (Interviews 5, 8). Regardless of the negative image, the activists have worked together with some representatives of governmental bodies such as the mayor of Petorca District (Interviews 6, 8). Gabriel (Interview 5) states that he has contacted various governmental bodies from local to national level in order to report illegal water extraction.

Although, the activists have communicated and even cooperated with some governmental authorities, solidary motives cannot be found. This means, they did not cooperate because they wanted to add the profit of their counterpart to their own (as in $U_x = X + Y$). Rather, they concentrate on their own gains. I can clarify that by the example of the cooperation between Gabriel and Eva[33], a right-wing congress woman. Gabriel reports that he and everyone who reported water robbery was excluded from a re-

---

33   Name is fictitious.

union with the minister of Public Works (Ministro de Obras Públicas) at the local government (gobernación). He then was invited by Eva, who belongs to the opposition *»No estoy de acuerdo con eso, pero hay que aprovecharlo,«(»I do not agree but you need to use it.« translated by author),* says Gabriel and justifies his cooperation with her insofar as it enabled him to speak up for the problems in the village:

> »Èl que primero pidió la palabra y lo pidió que me lo diera a mí, era la Eva, hablé yo, y le dije, todo lo que tenía que decir, de la mesa técnica, de no ser invitado, y sale la mamá del gobernador, la Natalia Esteban que es concejal aquí a decir, me interrumpió, como defendiendo al gobernador, a decirme, que todos los presidentes de canales no quisieron que yo fuera, y habían dos presidentes que levantaron la mano, »yo nunca dije eso«, dos la dijeron a la cara, o sea quedó como ridícula.« (Interview 5)

> (»The one who first asked for the floor and who asked to give it to me, Eva, I talked and I said everything I needed to say, about the water round table, about not being invited, and there goes the mother of the governor, Natalia Esteban who is [municipal] councilor here and says, she interrupted me, she defended the governor and tells me that all presidents of the WUO [canalistas] did not want me to come, and then two presidents raised their hands: 'I never said that', two told her that in her face, in other words, she made herself ridiculous.« Translated by author)

This statement also shows his experience of rejection by the local government.

To sum up, the activists have a profoundly negative image of the *public institutions*. They cooperate out of self-interest focusing primarily on the egos benefit ($U_x = X$). Hence, their orientation can be described as selfish-rational, respectively hostile in the case of Pablo. Only Mateo has a different point of view. He underlines his cooperation with various governmental authorities. He does not want to blame governmental bodies or their representatives because he believes that they do not have the legal and human resources to fulfill their work adequately. He believes that governmental bodies need to be part of the conflict solution (Interview 7). In that sense, his orientation of interaction with governmental authorities is cooperative ($U_x = X + Y$).

The orientation of interaction towards *civil society* is classified as cooperative. Since all activists strongly identify with farming as well as with their province and believe that people should organize more, they try to approach *civil society* and underline their cooperative relationship. Pablo (Interview 8) says that motivating and uniting people is not an easy but important task. Also, Juan reports having difficulties convincing his community members – especially the younger generations who he feels occasionally disrespected by. However, he believes that future generations need to care more about their environment and therefore wants them to engage with MODATIMA. Mateo complains that the group has never received strong support from society and small-scale farmers (Interview 7). Despite their criticisms towards civil society, the analysis shows that the activists consider the gain of the community as their own gain. Hence, the orientation of interaction is classified as cooperative ($U_x = X + Y$).

The same cooperative behavior was observed towards their own activists' group. Despite certain internal issues about process strategy and public affairs, activists recognize gain of another activist's member as a gain for themselves or the whole group. Mateo pulled out of MODATIMA because he disagrees with the aggressive image some of its members give to the movement. According to him, it is important to fight in a non-violent and face-to-face way. Additionally, he believes that some people have joined the movement for opportunistic reasons. Unfortunately, he does not explain this claim in detail (Interview 7). In sum, the orientation of interaction towards members of their own actor group is mainly cooperative ($U_x = X + Y$).

7.5.1.3 Citizens

In addition to the *activists* and *local leaders*, I decided to analyze four more interviews with persons who I classified as *citizens* of Petorca Province. These interview partners are members of WUOs. However, they do not have a leading role, neither in that WUO nor in other organizations. Therefore, they represent a wider range of citizens that are not as visible as other actors in decision making processes. Nevertheless, they are probably the largest actor group in terms of numbers. They are used as a reference and legitimation in decision making processes as *local leaders*, *activists* and politicians claim to work for their benefit. I defined this actor group as aggregated, because the *citizens* do not represent any organization, but they relate to each other on a geographical and social level. The following Table 12 shows some details about the interviewed persons. Additionally, I analyzed observations made in WUO meetings in this subchapter.

Table 12: Citizens. (Own research)

No.	Names (fictitious)	Sex	Age	District	Profession
9.	Paula	F	50-65	Cabildo	Unemployed
10.	Sofía	F	65-80	Petorca	Small-scale farmer
	David	M	65-80	Petorca	Pensioned, small-scale farmer
11.	Inés	F	65-80	La Ligua	Pensioned, small-scale farmer
	Vera	F	50-65	La Ligua	Small-scale farmer
12.	Iván	M	50-65	La Ligua	Small-scale farmer

OVERVIEW

All participants evaluated in this actor group are personally affected by the water scarcity. While one of them gave up farming, the others have tried to adapt to the water scarcity by technical improvements and crop changes. The interview partners show a strong connection with the traditions of farming in their families. Therefore, they point out concerns about the loss of small-scale agriculture caused by the water scarcity in Petorca. They believe that both, drought and an unfair system of water distribution and access, are responsible for this development. In specific, they criticize that low monitoring of water use, corruption and water privatization reinforce the water conflict. Nevertheless, their perception of governmental programs is not generally bad. Almost half of the *citizens* say that they are satisfied with the local governmental support. Furthermore, just as the other subgroups of *society*, the *citizens* observe a lack of cooperation and describe a divided and individualistic society. The orientation of interaction towards different actor groups is marked by selfish-rational behavior. Towards *large-scale agriculture,* a competitive orientation can be ascertained. In the following part, the codes found in the interviews and their interconnections are explained in detail.

MOTIVATIONAL ASPECTS

All interviewed citizens report to be personally affected by water scarcity. Four interview partners received potable water by trucks in the past (Interviews 10, 11). But also irrigation water became scarce. Iván reports that his family almost lost all their income as his father, an owner of a small-scale stockbreeding business, was forced to reduce 90 % of his business (Interview 12). Others suffered economic loss because of investments in irrigation systems (Interview 10) and buying additional irrigation water (Interview 11). Facing that situation, the farmers decided to change their strategies. Inés and Vera changed from bean and potato crops to flowers, which require less water, as early as nine years ago. Further, they built a new well which served them for about three years. Later, they changed to a technical irrigation system (Interview 11). David and Sofía are also about to change to that irrigation system. Further, instead of beans, they plan to cultivate green vegetables with the help of the governmental program INDAP. Former changes to walnut and olive trees were not successful because of the little water availability (Interview 10). Iván changed the family business to quinoa when he joined the farmers cooperation Petor-Quinoa in 2014 (Interview 12). Paula is the only one who decided not to continue her family's farming tradition. She inherited a part of the six hectares land of her mothers who used to cultivate potatoes, corn and lentils. Traditionally, her family irrigated with canal water. Recently, there is no water in the canal and the only option left, according to Paula, is to install a well. The costs for the installing and maintenance of the well together with the large distance to the land owned made her lose interest in farming (Interview 9).

Except for Paula, the motivation of the interview partners to change their farming strategies is related to their identities. All interview partners come from families of small-scale farmers. For example, Iván, who has an academic background in engineering, underlines his decision to save his father's farming business when his father was about to sell his land (Interview 12). David stresses the advantages of self-sufficient farming to become independent from markets (Interview 10).

Politically, none of the interview partners expressed his or her identification with a certain ideology or party. Paula (Interview 9) says that although she likes to follow political activities in the news, she does not want to be part of that. Iván claims to be politically independent and points out his values of equality (Interview 12). Moreover, David emphasizes rule-consistent behavior and complains about being excluded from the WUO's reunions after requesting payment and rule compliance (Interview 10).

COGNITIVE ASPECTS

In addition to the effects of water scarcity on their personal life, the interview partners point out negative impacts on society. Mostly, they worry about the loss of small-scale and self-sufficient agriculture (Interviews 9-11). The interview partners observe that people sell their land and water rights (Interview 12) and start to work for bigger agricultural companies (Interview 11), change to other working areas or leave Petorca Province (Interviews 9, 10, 12). Furthermore, Iván (Interview 12) points out the loss of biodiversity. Sofía and Iván underline that the water scarcity impacts harm small-scale farmers only (Interviews 9, 10). For this reason, Iván does not believe that the situation is caused by a natural drought. He says:

»[...] o sea si fuese sequia todos estaríamos complicados, es una cuestión lógica esa yo creo. Pero eso no se ve, solamente el que está más afectado es el más chico.« (Interview 9)

(»[...] well, if it would be a drought, all of us would be in trouble, this is logical, I believe. But you don´t see that, the only ones who are really affected are the small ones.« Translated by author)

Iván states that the water scarcity roots in distributional problems. The other interview partners have a similar point of view. Nevertheless, they recognize a severe precipitation decrease and see climate change as a reason (Interview 9-11) – except Iván (Interview 12) who doubts its existence. They believe that the unfair system of water access and distribution favors large-scale agricultural companies. They stress that this system is the driving factor of the conflictive situation (Interviews 9, 10, 12). The citizens clarify this conclusion with examples from their own experience as some report to depend on water that is left over by neighboring large-scale agricultural companies and underline that those companies are subsidized by the state (cf. Interviews 10, 11). What aggravates this perception is their suspicions of corruption and conflicts of interest (cf. Interview 12).[34] For example, small -scale farmer David states:

»Los políticos son muy vendidos. Hay grandes y todos los grandes entre ellos se arreglan. Corrupción.« (Interview 10)

(»The politicians are sold. They are big and the big ones sort things out between themselves. Corruption.« Translated by author).

Therefore, he sees a monitoring system organized by the users such as the Junta de Vigilancia as the only solution for the scarcity (Interview 10). Iván (Interview 12) reports about conflicts of interest. He complains that the law is heavily influenced by large-scale agricultural business and consequently serves the interests of a few instead of assuring equality:

»El marco regulatorio es bastante ambiguo por decirlo menos [...], en el caso de acá de la provincia, las mismas personas que hacen las leyes son grandes productores [...], o sea tienen el mango por el sartén, como es el dicho ese, ellos tienen la ley, tienen el poder y tienen todo.« (Interview 12)

(»The regulatory framework is pretty ambiguous, at least, [...] here in this province the same persons who make the law are big producers [...], in other words, they are pulling the strings, they own the law, they own the power and they own everything.« Translated by author.)

---

34 It needs to be mentioned that Vera confuses corruption with her perception of an unfair subsidy system (Interview 11).

Hence, the interview partners perceive that the water management system favors large-scale agricultural companies and that large-scale agricultural companies have a high influence on the political system. They see the expansion of large-scale agricultural companies in Petorca as culpable for the water scarcity. For example, some interviewees describe how the increase of well constructions by large-scale agricultural companies upstream diminishes the water that reaches their land (Interviews 10, 11). Additionally, some interview partners believe that the water extraction of large-scale farmers is higher than their actual water rights and accuse the agricultural companies of water robbery (Interview 10). Iván also includes small-scale farmers and calls it an »open secret« that everyone extracts water illegally (Interview 12).

Those accusations of illegal water extractions relate to the citizens' complaints about lack of monitoring and compliance assurance (Interview 10-12; Observation, December 13, 2016). For instance, in a WUO meeting a small-scale farmer complained about an illegal water infrastructure of his neighbor. As a lawyer recommended to inform the police, he laughs and replies: »*Pero estamos en Chile, mijia*« (»*But we are in Chile, my dear.*« Translated by the author; Observation, April 8, 2016) This occasion indicates the low trust in governmental bodies and authorities.

Further, the interview partners express their dislike of the privatized water system. Vera especially criticizes the opportunity to change the extraction points of the water rights (cf. chapter 6.4). In addition, both, Vera and Iván complain that water has a higher monetary value than land. They believe that water should be treated under a different regulatory system which recognizes the resource as a common good (Interviews 11, 12).

Only Paula stands out with her opinion about the water management system. On the one hand, she recognizes that under the given conditions people with lower financial resources as herself have more difficulties to access water. On the other hand, she insists that the system needs to be maintained like it is because changes would entail even more costs. She concludes: »*Entonces, hay que dejar que los grandes siguen para arriba y nosotros seguimos atrás, aunque nos cueste.*« (Interview 9) (»*So, we just have to let the big ones move forward and we are following behind, even though it's hard for us*« Translated by author). According to her, in the near term, the cleaning of the canal and, in the long term, rain are the only solutions to water scarcity (Interview 9).

The opinions about governmental programs to improve water management are diverse. Sofía and David are satisfied with the support they receive from the governmental program for small-scale farmers

PRODESAL (Interview 10) and Paula stresses that the different govern-mental bodies perform correctly and feels empowered by the workshops they offer (Interview 9). Inés and Vera, however, do not believe that solu-tions offered by the government, such as water tanks, will actually be im-plemented. They feel overruled by the last reform that urged them to change their water rights' legal classification from *shares* to *liters per sec-ond* because they feel they have been not been well informed (Interview 11). Iván complains about the injustice in governmental solution programs because the canal of his community was left out of a bigger construction project. According to him, his WUO is too small to produce enough polit-ical pressure and therefore is not taken into account by the government (Interview 12). Further, in WUO meetings small-scale farmers complain about INDAP. They criticize that the program does not fund the projects that are indeed needed (Observation, April 8, 2016), and that it is too cost intensive to enter the program (Observation, December 13, 2016).

With regard to their own communities, the interview partners witness low interest in cooperation. They say that only a few people participate in meetings and no one is willing to take over a leading position (Interviews 10, 11). Vera and Ivan report that people are not interested in agriculture anymore because they have lost hope of being able to live on their small-scale productions. They underline that people are not united and prefer in-dividual solutions (Interviews 11, 12). Further, David complains that he had a hard time when he was a WUO leader because the members did not want to contribute financially (Interview 10). Iván tells about the difficul-ties he experienced when he tried to organize politically (Interview 12). He underlines that he and other small-scale farmers do not have the time to participate in social or political meetings because their work is too de-manding. Similar perceptions were observed in different WUO meetings. Members complained about the lack of participation (Observation, April 6. and 13, 2016) and reported that often individualistic behavior restricts people from cooperating (Observations, December 13, 2016)

ORIENTATION OF INTERACTION
As explained in the cognitive aspects of this actor group, all interview partners – except for Paula – blame large-scale agriculture industry for the water scarcity. Most interviewees compete directly with large-scale farm-ers as they share the same water source (Interviews 10-12). David believes that these companies are unscrupulous insofar as they would not care about leaving poor people without water. Despite one confrontation with a manager of a neighboring company, David declares to have no contact

with large-scale agricultural companies (Interview 10). Unlike David, Iván believes to have good access to different companies and states that he could even discuss with their representatives about opinions on water scarcity. However, these talks are restricted to informal occasional meetings, (Iván 74) and he believes that his criticism is not taken seriously:

> O sea, claro, tú le puedes plantear el tema, pero es como David y Goliat, te van a escuchar y si lo pueden hacer, dependiendo con quien tu hables de la empresa, o te va a escuchar o simplemente te va a decir que estas equivocado y tu como chico no es mucho más lo que vas a hacer. (Interivew 12)

> (»Well, of course, you can talk about the issues, but it's like David and Goliath, they will listen and if they can, depending on who of the company you are talking to, whether he will listen, or he will just tell you that you are wrong and you as a small person, there is not a lot left to do.« Translated by author)

The qualitative data indicate that interaction orientation of *citizens* towards *large-scale agriculture* is mainly competitive ($U_x = X - Y$). However, Paula's orientation is different from the rest of the group. She underlines having a neutral relationship with large-scale agricultural companies (Interview 9). Thus, her orientation of interaction towards *large-scale agriculture* is identified as selfish-rational ($U_x = X$).

Another actor group mentioned by the *citizens* is *governmental authorities*. Except for Inés and Vera, all interview partners claim to cooperate with governmental bodies. However, considering their negative perception of governmental programs and the water management system, this cooperation is not based on solidary motivation. Rather than that, the interview partners are driven by their motivation to maintain their (families) farming traditions. Further, Iván points out that he confronted governmental authorities with his criticism. For instance, he reports about a discussion with some representatives of the DGA and DOH. He complained to them about the lack of a functioning inner-communication and criticized that they cause bureaucracy and transaction costs for the farmers (Interview 12). Although the actor group has a rather negative perception of *governmental bodies and authorities*, they cooperate primarily with regard to their own gain. In conclusion, I classify the orientation of interaction towards *governmental authorities* as selfish-rational ($U_x = X$).

With regard to their own actor group, *civil society*, a selfish-rational ($U_x = X$) interaction orientation dominates also. The interview partners underline the individualistic behavior and the lack of cooperation among each other. Only Iván reports that he cooperates with his family and neighbors in order to share water or to support each other's harvest (Interview 12).

Hence, some cooperative orientation of interaction ($U_x = X + Y$) is registered towards *civil society,* too.

The same selfish-rational interaction orientation is observed towards *activists*. Although at least three of the interview partners support the ideas of the *activists* (Interview 11, 12), none of them has participated in their activities. While Iván says that he has no time for political activism, Inés and Vera declare, that they are afraid of participating because of the potential violence that goes along with activism:

> »Somos cobardes no más. La otra vez también hubo una protesta porque iba a llegar una minera en las dunas y fueron a protestar para allá, después con un chorro de agua botaron a una señora creo unos carabineros, así que no es muy bueno ir a eso. (Interview 11)

> (»We are just cowards. The other day there was a protest because of a mining company in the dunes and they went to protest, and then, with a water stream the police threw down a lady, I believe. That is why it is not good to go there. Translated by author)

In other words, the interview partners show solidarity with the activists' motives but would not actively express their solidarity. Accordingly, I classify their orientation of interaction towards activists as selfish-rational ($U_x = X$).

7.5.2 Actor group: governmental bodies and authorities

As demonstrated in chapter 7.4, a large number of Chilean governmental bodies influence water governance directly or indirectly. In order to capture a broad range of governmental bodies, I divided them into three groups with regard to their spatial area of responsibility. Further, the selected interview partners are a mixture of civil servants and elected authorities, respectively authorities with positions that are selected by the current government. I chose this classification as the members of each actor group share responsibility or take over tasks on a similar political level. Moreover, they act in a hierarchical structure due to their positions in public institutions. As representatives of the state, their actions should be highly independent from their personal benefits. Consequently, I classified them as cooperative actors. In addition to the interviews, field observations were considered in this analysis.

7.5.2.1 Local authorities

The community level is considered as the spatial range of the subgroup *local authorities*. Thus, the actor group consists of municipal civil servants and workers of the governmental program for small-scale farmers PRODESAL as well as elected authorities. The following Table 13 shows the interview details:

Table 13: Local authorities. (Own research)

No.	Names (fictitious)	Sex	Age	District	Profession / Position
13	Hugo	M	65-80	Cabildo	Elected authority
14	Irene	F	20-35	Petorca	Geographer, civil servant at municipality of Petorca
15	Marta	F	20-35	La Ligua	Agronomist, civil servant at PRODESAL
16	Mario	M	50-65	La Ligua	Elected authority

OVERVIEW

In short, the data indicate that the participants feel a strong connection with their community and with the rural area. While half of them report to be personally affected by the water scarcity, everyone reports about negative consequences for society, such as the impoverishment of small-scale farmers. In addition, they mention the loss of culture and tradition. Two of the interview partners claim to be of left-wing political orientation. Like the actor group *society*, they regard climate change and expansion of large-scale agricultural companies as main reasons for the water scarcity. They mention that the number of governmental bodies is too diverse and therefore produces high bureaucratic obstacles and low transparency. This issue, together with a lack of water regulations and monitoring, incites illegal water extraction and corruption, according to the interview partners. In addition, they blame the water privatization for provoking unequal water distribution and access. Moreover, they complain that society neither organizes cooperation, nor participates in activists' events. The interview partners believe that a lack of environmental awareness and the fear of social exclusion and job losses restrain the cooperation within society. The interaction orientations are predominantly cooperative or selfish-rational towards the actor groups *civil society*, *activists* and other *governmental authorities*. Towards the latter and towards *large-scale agriculture* also competitive behavior was observed.

MOTIVATIONAL ASPECTS
The interview partners' identities are based on diverse values and motivations. Three of the interview partners consider themselves as political persons. Hugo is politically independent, and Mario belongs to a left-wing party (Interviews 13, 16). Irene shows her political identity not only in her left-wing political beliefs and her background as an environmental activist but also in her function as a(n informal) consultant of the mayor of Petorca (Interview 14). Both, Irene and Mario consider equality as their core value (Interviews 14, 16).

All interview partners state to have a strong relationship with their community, especially within the rural area. For instance, Hugo underlines repeatedly that he was born and raised in Cabildo and therefore knows the area very well. Irene decided to live in Petorca to be closer to the community. She repeatedly stresses her dedication to the community. Her decision to live in Cabildo and her voluntary work in the community underline her empathy (Interviews 13, 14). Hugo and Marta seem to have a strong connection to small-scale agriculture. Hugo frequently says »us« while referring to small-scale agriculture and points out that he cultivates land himself. For example, he says: *»[...] y por lo tanto somos los chicos que teníamos como 5 hectáreas, vivimos con desazón.« (Interview 13) (»[...] therefore, it's us, the small ones who own 5 hectares, we live with anxiety« translated by author).*

The dedication and closeness of Marta and Irene towards their clients is also revealed by their dissatisfaction about the water conflict. While Irene expresses her feelings literally as frustration and pain, Marta reveals her feelings indirectly, for instance, by ironically considering »to kill everyone« as the solution to the problem (Interviews 14, 15). Hugo and Irene are personally affected by the water scarcity. Hugo has lost crops and Irene has struggled with access to potable water (Interviews 13, 14).

COGNITIVE ASPECTS

The water scarcity impacts mentioned by the interview partners are mostly interdependent. They are rooted in the economic damage of agriculture that causes poverty rise and loss of small-scale farming (Interviews 14-16). Further effects are unemployment rise, loss of traditions, selling of land and concentration of land owned by private persons (Interviews 14, 16). Moreover, the interviewees report about problems with the drinking water supply by water trucks (Interviews 14, 16) and worries about the water quality (Interview 13).

All interview partners of this actor group agree on two main factors that cause water scarcity in Petorca Province: climate change and expansion of agriculture (Interviews 13-16). All of them report about illegal water extraction by large-scale agricultural companies (Interviews 13-16). They argue that this behavior harms the environment (Interview 14) and the most vulnerable parts of society, namely people living in rural areas and small-scale farmers (Interview 13-16). Consequently, they perceive water distribution and access as unfair. Especially Hugo and Mario criticize the water privatization and claim that the resource should either be nationalized (Interview 16) or, as Hugo argues, the regulations should be changed to correspond to the status before the water code of 1981 (Interview 13). Marta does not have such strong ideas about the water regulation. Nevertheless, she also mentions nationalization as a possible solution. In first place, she and Irene demand prioritization of drinking water use (Interviews 14, 15). Irene, who works closely with committees of community water providers (APR), emphasizes to treat water as a common good and organize community water management (Interview 14).

The criticism of this actor group is not only restricted to the juridical water regulation but also concerns its implementation. The actors complain about the high number of governmental bodies and state that they do not coordinate well because of their competitive behavior, the slow and non-transparent processes, and the distance from society. Hugo concludes: »*Son como primos, pero no se entienden entre ellos, no hay caso.*« *(Inter-*

view 13) (»*They are like cousins, but they don't get along with each other.
There is no hope.*« translated by author) The problem of the diverse governmental bodies has been recognized by the province's governor as well.
In a community meeting, he stated that the water management took a reactive attitude because it was only reacting to the damages of water scarcity
instead of actively preventing it. He suggested uniting the governmental
bodies in one water management office for Petorca Province (Observation,
November 30, 2016). Mario especially criticizes the governmental bodies
DGA and DOH for having no sufficient financial and manpower (Interview 16). According to the interview partners, this situation has led to a
lack of water use monitoring and sanctioning which favors illegal water
extraction (Interviews 13, 14, 16). Further, they state that the municipalities mutually reported water robbery, but they were never successful in
court. Irene blames the DGA for oversharing water rights. She demands:

> »Sanciones reales, duras al robo del agua, al crimen ambiental en general, porque
> se hace, hay mucho desastre ambiental en Chile y no hay penas, que de verdad que
> valgan la pena. Sino que simplemente se transforman en costos de la
> oportunidad.« (Interview 14)

> (»Real and hard sanctions for water robbery and environmental crime in general,
> because that is what happens, there is a lot environmental disaster in Chile and no
> penalty, which really count as penalty. Instead, the penalties just turn into oppor-
> tunity costs.« translated by author)

Despite the lack of effective sanctioning, Irene worries about security for
those who monitor law compliance. She reports about civil servants who
received threats and lost their jobs because they had investigated water
robbery (Interview 14).

Furthermore, all interview partners suspect corruption or conflict of interest in water management. In specific, they are concerned about the relationship between owners of large-scale agricultural companies and governmental authorities. Hugo says that a lot of politicians are large-scale
landowners themselves. Thus, to assure their own benefits, politicians will
not change the law. Moreover, he believes that civil servants pass sensitive
data to entrepreneurs. While he suspects that corruption exists on a higher
level, he believes that there is not such crime in Cabildo (Interview 13).
Mario is rather reluctant to talk about that topic. He confirms that illegal
water extraction could be related to corruption. Nevertheless, he points to
the judges who must decide on this and explains that he is not allowed to
mention any corruption if it has not been verified by the court. He refuses
to discuss this topic further because of the costs that could be inherent to it
(Interview 16). To understand his rejection of this topic, it is crucial to

mention that he has been accused of corruption himself. For example, in 2016, his webpage was hacked so that users see a screen with information about his corruption activities. Furthermore, during informal meetings with community members, activists and others, different cases were addressed accusing him of misuse of power and conflict of interest (field observations). Contrary to Hugo and Mario, who think corruption is a problem on a national but not local level, the civil servants report about corruption and conflict of interest on the community level. Moreover, they criticize that governmental programs heavily depend on the interest and the skills of the civil servant in charge. Marta emphasizes that people are disappointed about the offered programs as they seem to come to no end. Irene is specifically critical about the water tanks that, according to her, mostly serve large-scale agricultural companies (Interviews 14, 15).

Another topic stressed by all interview partners is the lack of cooperation within the community. Hugo and Mario complain about the lack of participation in community organizations and the little awareness of the water conflict from people who are not directly affected (Interviews 13, 16). Mario argues that people are opportunistic because they only cooperate when they find themselves economically affected. As soon as their personal problem is solved, they lose interest. Also, he believes that Chilean people are traumatized by the dictatorship. Firstly, they do not trust anybody – not even their neighbors. Secondly, they do not want to be associated with the political left and consider it *»como si fuera lo peor de lo malo que hay en esa sociedad.«* *(»as the worst of the worst that exists in this society«)* (Interview 16). However, this rejection of politics is not restricted to one side of political ideology. The right-wing governor stated in a public meeting that politics suffered of mistrust by the community and emphasized a strengthening of the commitment of both society and government in the future by implementing more horizontal management (Observation, November 30, 2016).

Marta and Irene find further reasons for the lack of cooperation. Irene says that people in Petorca Province have lost hope because they were promised help that never arrived, and they feel powerless confronting large-scale agricultural companies and their political network. Marta reports about a lack of community leaders. She states that as the small-scale agricultures have no interest in participating in meetings as they do not find any achievable solution. Additionally, she says that they are mostly elderly people with low educational background. She concludes that the lack of education complicates WUOs' management and that their members are vulnerable to frauds. She explains that once a person has gained their trust: *»firman las letras que no pueden leer«* (Interview 15) *(»they sign letters that they can't read«* translated by author). Further, Irene says that people are afraid of getting involved in water management while working for large-scale agricultural companies. Irene stresses that APRs are an exception from other WUOs because they cooperate well with each other. Nevertheless, she experiences rivalry, particularly between APRs and ACs[35]. She believes that APRs are more vulnerable than ACs because the latter have more political power (Interviews 14, 16). Because of this lack of interest in cooperating, Hugo and Mario consider social and environmental movements as weak in terms of participant numbers (Interviews 13, 16). Marta underlines that people are tired of listening to *activists* that report about the same problems over and over. She does not see any positive outcome of the activists' actions, be it demonstrations or signature collections (Interview 15).

Regarding conflict resolution, Mario points to the technical solutions. He strongly promotes the idea of a desalinization plant for potable water to supply the districts of La Ligua, Cabildo and Petorca. Hugo supports Mario's plans to alleviate water stress by creating bigger supply. Additionally, Mario emphasizes the cultivation of different crops as part of the solution (Interview s13, 16). Furthermore, all interviewed persons believe that a water code change must be part of the conflict's solution. They suggest that the Chilean government should either nationalize water (Interviews 13, 15, 16) or assure a community-based management (Interview 14). Irene further demands higher sanctions on water robbery and environmental mismanagement in general (ibid.). According to Hugo, the key for achieving a water code change is political will, but he believes that especially because of the strong connection between the agricultural business and the congress members and based his own experience of working with the congress, such political will for change does not exist (Interview 13). Irene and Marta also point to a rise of political and environmental awareness in order to achieve change (Interviews 14, 15). Mario underlines the municipality's ambition to raise political awareness by informing communities. Nevertheless, he remains disappointed about the outcome as people still do not organize or vote differently (Interview 16). Contrary to the other interview partners, Hugo hopes for a natural solution to the water conflict. Throughout the interview he repeatedly expresses his relief about

---

35 For more information about different WUO please read chapter 7.4

the rains of the ongoing year and his hope that precipitation will last and alleviate the conflict (Interview 13).

ORIENTATION OF INTERACTION

Almost all interview partners claim to have no or little contact with *large-scale agriculture*. They believe that the owners of large-scale agriculture are unscrupulous (Interviews 14, 15) and reported illegal water extraction by their companies (Interviews 13, 16). Because their aims with regard to water governance compete, I classify the orientation of interaction of *local authorities* towards *large-scale agriculture* as competitive ($U_x = X - Y$).

With regard to other *governmental bodies and authorities*, the interaction orientations differ depending on the regional scale. Orientation of interaction towards local governmental bodies differ between cooperation and competition. Cooperation, especially towards and within the municipality, is the predominant interaction orientation for all interview partners. However, the civil servants also report about downfalls of that cooperation. For instance, Irene sees WUOs, such as ACs, as competitive to her clients and Marta feels like a marionette of higher institutions (Interviews 14, 15). The data confirm that the orientation of interaction towards higher governmental bodies is rather selfish-rational. Their goals in water governance seems to be impeded by decisions made on a higher political level. For instance, Hugo and Mario complain especially about the national congress (Interviews 13, 16). Thus, the orientation of interaction towards other governmental authorities differs between mainly cooperative ($U_x = X + Y$) towards the same regional scale and competitive with regard to higher *governmental bodies and authorities* ($U_x = X - Y$).

With regard to *civil society*, all interview partners show a cooperative orientation of interaction. While the Hugo and Mario base their interaction on listening to and informing the communities, the cooperation of Irene and Marta seems to reach further as both express their trust in communities and are trying to find conflict solutions mutually (Interview 14, 15). Hence, in general the orientation of interaction is classified as cooperative ($U_x = X + Y$).

Another actor group mentioned by the local authorities is *activists*. None of the interview partners reveals a negative image of them. Everyone claims to either support *activists* or show selfish-rational orientation insofar as they sympathize with their ideas but not with their way of acting. Thus, the interaction orientation towards *activists* differs between cooperative ($U_x = X + Y$) and selfish-rational ($U_x = X$).

7.5.2.2 Regional authorities

The interviews analyzed in this chapter were held with members of governmental bodies that directly or indirectly influence water governance and represent the Vth region of Valparaíso. Ramón works in a leading position at the regional Ministry of Agriculture, and Roberto works in a leading position at the regional Ministry of Public Works (MOP). The third interview partner, Álvaro, is a civil servant of the National Irrigation Commission (CNR) that belongs to the Ministry of Agriculture (cf. Table 14).

Table 14: Regional authorities. (Own research)

No.	Name (fictitious)	Sex	Age	Region	Profession / Position
17	Ramón	M	50-64	Valparaíso	Regional Authority
18	Roberto	M	50-64	Valparaíso	Regional Authority
19	Álvaro	M	35-50	Valparaíso	Regional Authority

OVERVIEW

Ramón and Roberto declare social equality as their core value and identify with the political left. Ramón repeatedly underlines his dedication towards small-scale agriculture. Álvaro prefers not to talk about his personal and political values. Basically, the interview partners name four main reasons for water scarcity in Petorca Province: lack of technical water infrastructure, lack of community organization, climate change and water regulation. However, they have different focal points. For example, while Ramón questions climate change and believes that direct governmental help is more important than water regulation reforms, Álvaro does not see a problem in the current water regulation at all. The predominant interaction orientations of the interview partners towards other actor groups are selfish-rational and cooperative. The following analysis demonstrates further details.

MOTIVATIONAL ASPECTS

While the civil servant Álvaro claims to belong to no political party and refuses to speak about his personal or political values, Ramón and Roberto reveal a few details about their identity. The core values determining Roberto's work are equality, social justice, and diversity (Interview 18). Whereas Ramón agrees with equality and social justice, he gets more specific by underlining the support of small-scale farmers. During the interview, he repeatedly mentions his closeness to small-scale farmers by stressing his care for their well-being and his dedication to work on-site. Also, he seems to enjoy his work as a coordinator and manager and likes to offer solutions for the community insofar as he sees himself as a nexus between different governmental bodies and civil society. By repeating his willingness to »lead the way« and »open up opportunities« to small-scale farmers, he shows that he is proud of his work and underlines that his work is gratefully appreciated by the community. The following example illustrates his self-perception as an outstanding support for the community:

> »Muchas veces me han dicho: Usted no más contesta. Los demás no contestan. Bueno yo les ayudo. Yo les hago escuchar. Yo les hago las reuniones. Eso puedo hacer.« (Interview 17)

> (»A lot of times they say: 'You are the only one who replies. The others don´t reply.' Well, and I help them. I make them heard. I make them having meetings. That is what I can do.« Translated by author)

Moreover, he underlines his long-term working connection with both the Petorca Province and the field of agriculture in which he graduated and mentions environmental protection as a core value of his work (Interview 17).

COGNITIVE ASPECTS

The interview partners give different answers to the question about the water scarcity's reasons. Álvaro believes that the state lacks financial power which blocks the building of effective water infrastructure in the area. He underlines that technical solutions exist. However, in Petorca Province, he claims, people neither cooperate to push through those innovations, nor work jointly on other solutions. According to him, community members do not want to take on the responsibility for their irrigation systems (such as cleaning their canals) and lack trust towards each other. He emphasizes that *civil society* must stop being lazy and actively take charge of the situation:

> »[...] hay un tema en que en nuestra sociedad, se tiene que sacudir la comodidad, el ser cómodo y no querer participar y que alguien más haga el esfuerzo, eso hay que sacárselo de encima. Los cambios no se logran estando sentados esperando que el vecino haga las cosas, haga algo por mí ¿te fijas?« (Interview 19)

> (»[...] there is this issue in our society, it must shake off the comfort [attitude], this being lazy and not willing to participate and wanting others to take care, they must get rid of this. No change will be achieved [if I am] waiting for the neighbor to do the work for me, you see?« translated by author)

Moreover, he claims that farmers in Petorca copy each other's behavior on how or what to cultivate without questioning it due to their lack of education (Interview 19).

According to Roberto, the main reasons for the scarcity are, firstly, climate change reinforced by the geographical conditions of Petorca Province (no natural snow accumulation) and, secondly, unequal water access rooted in the water code and nurtured by the lack of community organization. In specific, he criticizes that water privatization does not prioritize water for human consumption (Interview 18).

While Ramón recognizes drought as a structural change which was influenced by humans, he hesitates to call this phenomenon (anthropogenic) climate change and refers to recent scarcity alleviation due to increased rainfalls and farmers adaptation. Like Roberto, he believes that the conflict in Petorca roots in the legislative water regulation. In his view, under the current water code small-scale farmers are more vulnerable than others because they have less options to access water due to scarce financial and educational resources. In general, he claims that inequality in Chile roots in the nations policy system which follows a neoliberal ideology. According to him, this policy of a reduced state also explains the financial and legal weakness of the National Water Directive (DGA) and causes overexploitation and illegal water extractions. However, he believes that Petorca Province is not an exceptional case of illegal water extraction. Rather than that, he says that the region became stigmatized for being the province of water robbery. Likewise, Roberto criticizes the DGA´s low power and lack of water monitoring for causing water overexploitation and blames the institution for oversharing water rights (Interviews 17, 18).

Álvaro also recognizes overexploitation of water in Petorca. Contrary to the other interview partners, he does not believe that this situation roots in unequal access or a lack of regulation. In his opinion, the water code does not significantly influence the water scarcity in Petorca Province. He considers the existing regulation too restrictive and not able to adapt to current ecological and technological developments. Álvaro especially focuses on the defense of large-scale agriculture. He underlines that it is unfair that large-scale agricultural companies are blamed for the water scarcity. His first argument during the interview states that land owned by large-scale agricultural companies is not as much as suspected by *activists* and other criticizers and much less than the land owned by small-scale farmers.[36] He argues that large-scale agriculture is an engine for economic progress. Fur-

---

36 According to the data shown in chapter 7.1, this argument is wrong as approximately 90% of the cultivated land in the Vth region is owned by large-scale agricultural companies.

ther, he claims that there are bigger drivers of environmental degradation than agriculture in Petorca, such as the mining industry, and that agricultural companies have improved their ecological performance constantly, for example, by the efficient use of pesticides (Interview 19).

Ramón has a similar point of view on large-scale agriculture. He argues that it would be unfair to stigmatize agricultural entrepreneurs as bad people and states that large-scale agriculture brings progress to the province and that he has observed a growing environmental awareness of large and small-scale agriculture. However, he admits that it is more difficult for small-scale farmers to adapt to water scarcity. Contrary to that, Roberto believes that the environmental awareness of the community remains low. He does not only refer to the agricultural sector but detects a lack of awareness as a problem of Chilean society in general. Moreover, the Ramón and Roberto worry that awareness about the rural water scarcity is very low in urban areas (Interviews 17, 18).

With regard to solutions to the water scarcity, all interview partners point to technological measures. Álvaro mainly refers to technical irrigation systems and 'water highways'.[37] Roberto mentions inter alia new water sources, improved wells, water tanks and the re-use of recycled water. Further, Roberto and Ramón stress the importance of a change in the regulative system and demand to focus on social instead of economic issues. In specific, they refer to an improved protection of the right to water of small-scale agricultures. However, Ramón does not believe that a change in water regulation is the most efficient solution. Rather than that, he stresses the importance of public policies such as direct help programs which can be implemented faster and will have greater influence on the communities. Roberto emphasizes the need of improved water user organization, such as the implementation of a Junta de Vigilancia, which would ensure the monitoring of water distribution. Furthermore, he believes that education programs are necessary to cope with the water scarcity. This topic is also stressed by Álvaro who refers to the need of education for small-agricultural farmers and the importance of learning to adapt to technological and environmental change (Interviews 17-19).

ORIENTATION OF INTERACTION
Regional governmental authorities mention the following actor groups in their interviews: *civil society*, in specific *activists,* WOUs, and small-scale

---

37   For more information about those solutions please read chapter 7.1 [technologies]

farmers; other *governmental bodies* and *large-scale agriculture*. Starting with *civil society*, Ramón and Roberto declare to have a respectful relationship to the actor group of *activists*. Roberto says that they work on the same topics although they have different focal points. To progress in improving water infrastructure, he claims to need support from society. For that reason, he would like to strengthen his relationship with the *activists* (Interview 17). Hence, he supports their goals and shows cooperative ($U_x = X + Y$) interaction orientation.

Ramón's orientation of interaction towards *activists* is indifferent. He tells about occasional meetings with members of the environmental movement MODATIMA. However, he also reports that some of its members would refuse to talk to him and that the movements in Petorca are violent. On the one hand, he defends the people's right to protest, shows his understanding for the *activists'* actions and declares to have open doors for anyone. On the other hand, he underlines that conversations with members of MODATIMA are merely informal and that he will never have an official meeting with that group due to his position. At the same time, he complains that he has never been invited by them (Interview 17). Since his cooperative attitude seems to be more towards certain people than towards the *activists* as a group, I classify his interaction orientation towards *activists* as selfish-rational ($U_x = X$).

Álvaro does not mention activists' movements at all. However, it can be mentioned that his cognitive perception of the situation differs greatly from the perception of the local activists. Consequently, their aims in improving water management conflict (e.g. loosening regulations vs. strengthening regulations) and lead to a competitive interaction orientation ($U_x = X-Y$)

The next actor group mentioned by the interview partners belongs to *civil society:* members of WUOs and small-scale farmers. It should be said that all interview partners show a rather top down management approach when referring to the *civil society* and seem to be convinced of knowing how to lead the community. Ramón stresses his strong relationship and care for this actor group. He relates to have a close personal contact and states to build bridges towards other governmental bodies (Interview 17). Roberto also shows commitment to the community's needs. He underlines the participative strategy of his institution (Interview 18). However, those strategies basically seek acceptance of project development and focus on reducing resistance and explaining benefits, rather than opening space for self-organized bottom up structures. Álvaro does not mention any contact to civil society. However, his cognitive perception of the conflict shows neither hostile nor cooperative action orientation. Although he complains about people being irresponsible and inactive, he demands more education. Hence, the interaction orientation of the three interview partners differs. It is not clear if they consider gains of the civil society as their own gain, or if they rather focus on their own gain. Therefore, I classify the interaction orientation of all three interview partners as ranging between cooperative ($U_x = X + Y$) and selfish-rational ($U_x = X$).

Furthermore, all interview partners report to cooperate with other *governmental bodies* (Interviews 17-19). However, they do not show any strong orientation of interaction insofar as they neither complain nor underline the support of other institutions. Their cooperation with other governmental bodies seems to focus on the aims of their own institution. Therefore, I will classify their action orientation towards other governmental bodies as rather selfish-rational ($U_x = X$).

The last actor group mentioned is *large-scale agriculture*. Although Álvaro does not mention any direct contact to *large-scale agriculture*, he shows a strong connection to that actor group as he starts defending them at the very beginning of the interview before any question was ask. Especially Ramón and Álvaro underline the economic benefits of *large-scale agriculture*. Ramón underlines his good relationship with members of *large-scale agriculture* and that says that he cooperated with them in different meetings. Roberto only mentions *large-scale agriculture* once during the interview when he says that they have more capacities to adapt to water scarcity than small-scale farmers (Interview 17-19). Since particularly Álvaro and Ramón stress the benefits of large-scale agriculture companies for society, I will classify their interaction orientation towards this actor group as cooperative ($U_x = X + Y$).

7.5.2.2 National authorities

The actor group of national authorities consists of state's representatives taking on responsibilities on a national scale. The interviews were held with two representatives of the Chilean economic development agency CORFO (Eric and Carmen), a representative of the National Water Directive DGA (Carlos) and the elected congress member (diputada) Eva (cf. Table 15). Petorca Province belongs to her constituency and she is a member of the national congress's Commission of Water Resources (Comisión de Recursos Hídricos). The reader should know that the interview with Eva happened during a break in one of the commissions reunions and under the presence of further congress members. These circumstances have led to a rather short conversation. However, further attempts at a second interview failed. Her interview as well as the interview at CORFO was documented via written protocol. During the interview at CORFO, Eric led the conversation, while Carmen spoke very little.

Table 15: National authorities. (Own research)

No.	Name (fictitious)	Sex	Age	Profession / Position
20	Eric	M	35- 49	National authority
	Carmen	F	35-49	National authority
21	Eva	F	35-49	Congress woman
22	Carlos	M	50-64	Valparaíso

OVERVIEW

With regard to motivational aspects, Carlos shows dedication to his job and emphasis equality. Eva underlines her knowledge about Petorca Province and her preference of the former government of Sebastian Piñera, and Eric endorses his perspective on humans as win-maximizing beings. The interview partners provide various reasons for the water scarcity that are partly contrary to each other. While almost all interview partners argue that climate change is the main reason for the scarcity, Eric doubts its existence and calls the situation an outstandingly long drought. He believes that the water scarcity in Petorca Province is not as bad as people and media say and that small-scale farmers hinder solution strategies because they profit from the informality existing under the current water regulations. According to him, *large-scale agriculture* does not affect the scarcity. Carlos shows a contrary perception. He believes that *large-scale agriculture* and incentives for increasing agriculture by the state have led to water overexploitation. However, all interview partners agree on a problematic lack of users' organization caused by the mistrust inside society and towards governmental bodies. While Carlos believes in solutions such as a change in water regulation towards less neoliberal structures and prioritization of water for human consumption, Eric does not find failings in the current regulation. He suggests conducting studies as incentives for investments. Eva underlines the need for a law change towards stronger public institutions. Moreover, she demands education and training especially for small-scale farmers. With regard to interaction orientation, Eva stays indifferent which is why it is not possible to determine her interaction orientation. Eric and Carlos, on the other hand, seem to have opposing interaction orientations. While Eric shows a more cooperative orientation towards *large-scale agriculture* and a rather competitive orientation towards *activists* and *civil society*, Carlos expresses a cooperative orientation towards *activists* and *civil society* and a competitive orientation towards *large-scale agriculture*. His interaction orientation towards other *governmental bodies* remains indifferent with regard to all interview partners. In the following analysis, the motivational and cognitive aspects as well as the interaction orientation are demonstrated in detail.

MOTIVATIONAL ASPECTS

Carlos underlines his dedication to his jobs by proudly pointing out that he has worked at the DGA for more than 40 years. Although he concedes the existence of various issues of mismanagement by the DGA, he defends its work. Equality seems to be one of his core values as he repeatedly confirms his dedication to small-scale farmers and the right to potable water for everyone. Further, he shows interest for environmental protection and claims to have worked hard to achieve the introduction of an *ecological basin*[38] into the current water regulation (Interview 22).

Eric, Carmen, and Eva do not reveal much information about their personal identity or interest. However, Eva reports to have worked more than 10 years in Petorca Province and therefore, claims to know the area and water users well. Her identity shows closeness to the government of the former president Sebastian Piñera (2010-2014). Firstly, her party is member of his political coalition, and secondly, she emphasizes that the current government of Michelle Bachelet (2014-2017) is known to have worsened the situation in Petorca (Interview 21).

Eric highlights his view on humans as win-maximizing beings. This is important to the interview analysis insofar as this ideology of the classic

---

38   The introduction of an ecological basin is part of the current water reform debate. It aims at protecting a minimal amount of water of ecological threatened basins.

homo oeconomicos seems to serve as a base for his further argumentation with regard to reasons and solutions of the water conflict in Petorca (Interview 20).

COGNITIVE ASPECTS

In all interviews, climate change was mentioned as the main reason for the water scarcity. However, Eric doubts its existence. He complains angrily that he feels obliged to believe in climate change and suspects that it is primarily used as an excuse for other reasons, for examples, by scientists to receive research funds. Yet, he claims that the outstandingly long duration of the drought needs to be recognized and dealt with. Compared to the other interview partners, Eric's perception of the conflict stands out. He insists repeatedly that the situation in Petorca Province is not as bad as the media or people from Petorca claim (Interview 20).

Besides climate change, Carlos regards the win-maximizing national policy strategy as a driver for the conflict. He says that there is a lack of monitoring and sanctioning of water use while simultaneously incentives are given for augmenting plantations. According to him, this situation has led to the overexploitation of the resource. Carlos further criticizes the presence of too many governmental bodies that provoke organizational chaos. He includes his own employer (DGA) in his criticism and blames it for being financially and legally too weak to monitor and sanction water use properly (Interview 22). Eva and Eric agree on this and express the same concerns (Interview 20, 21). Yet, Eric believes that the expansion of large-scale agriculture does not have an impact on water scarcity (Interview 20). During the interview, Carlos receives a phone call which he later explains to be emblematic for the weakness of his institutions in terms of legal protection. He relates that he has been warned about an owner of a large-scale agriculture company who will try to sue him for reporting illegal water extraction. He explains that he gets no legal support by the DGA and must face this prosecution as a private person. Furthermore, Carlos criticizes the DGA for oversharing water rights by granting provisional water rights (Interview 22).

In the same line of argument, he expresses his dislike of the current regulation. In specific, he points out the water code and water privatization. In his view, water should be a common good and drinking water should be prioritized, whereas the current water regulation neither protects the right to water for human consumption nor assures water access for small-scale farmers. Moreover, he believes that the community organization and especially the trust in governmental programs and in society lack.

He finds reasons inter alia in the current organizational structures, such as WUO, which according to him are dominated by large-scale agricultural companies. As he states:

»¿Quién maneja el río? Él que tiene mayor cantidad de agua, por lo tanto, los presidentes de las juntas de vigilancias, asociaciones de canalistas, comunidades de agua, son solamente gente que tiene la mayor cantidad de agua. Nunca vas a ver al chiquitico, presidiendo, no es cierto, o gestionando el agua. Vas a ver al grande, al gran agricultor, no el chico.« (Carlos 109)

(»Who is managing the rivers? The one with the biggest share of water rights, hence, the presidents of Juntas de Vigilancias, AC and other WUOs are only people with huge numbers of water rights. You will never see a small person leading or managing water. You will see the big ones, the large-scale farmer, but not the small one.« Translated by author)

In addition, he suspects that some users regard the organizational chaos as beneficial insofar as the informal situation would allow them to extract more water than they are officially allowed to (Interview 22).

Likewise, Eric criticizes the cooperation lack between users and suspects the same. He states two hypotheses. First, the situation is not as bad as commonly acknowledged. Second, people do not see benefits in organizing as they are hiding illegal, respectively informal, actions. He reports that small-scale farmers often block the implementation of WUOs such as Junta de Vigilancia. Contrary to Carlos, he does not see a problem in the voting system that favors large-scale water right holders. He believes that people who oppose against the water system are underinformed. He does not see problems in power structures because he believes that, although the vote of users who own more water rights counts more (according to the number of water rights they hold), they would not be able to vote for or manage anything that is not in accordance with the existing law (Interview 20).

Eric regards states that all humans seek to maximize their benefits. According to him, this means that it is important to inform society about economic benefits of user´s organization. He underlines that the state and in particular the DGA do not have the financial resources to monitor, so that a users' cooperation is the only way to assure appropriate resource use. Carmen adds that this lack of cooperation roots in interpersonal mistrust which is part of the Chilean culture (Interview 20).

Eva gives some further reasons for the water conflict, namely unequal water access and water robbery, wrong investments, lack of national leadership, lack of awareness and migration into Petorca Province. According to her, solutions for the scarcity would imply improvement of water management with regard to the DGA and WUOs, and investments, especially in water tanks and river basin studies (Interview 21).

These proposals are in the same line as Eric's solution strategies. He demands further investigations on water availability to generate higher interest for public and private investments. He believes that studies showing the water resources availability will improve investment security (Interview 20). Carlos confirms these believes as he says that in some cases the DGA does not share water rights just because water availability remains unknown in certain areas (Interview 22) Moreover, Eric demands technical investments as he believes that a change in water regulations will not change the reality of life in Petorca (Interview 20).

Carlos argues differently. Although he concedes that investments in technical infrastructure, such as water tanks and technical monitoring systems, are needed, he insists that other technical solution such as improved irrigation systems are not that efficient. According to Carlos, technical solutions are not enough because they cannot compensate for the mismanagement in water governance. Therefore, he finds improved (and faster) public policies and an incremental change of the water regulation and constitutions towards more state power and less market-oriented structures necessary. In addition, he and Eva claim that there is a need for education and training especially for small-scale farmers (Interviews 21, 22).

ORIENTATION OF INTERACTION

The different actor groups mentioned by the interview partners are other *governmental authorities, activists, large-scale agriculture,* and *civil society.* Although, Carlos criticizes the multitude and the mismanaging of *governmental bodies*, he also reports about cooperation with the MOP and other ministries without complaining or showing any kind of hostile or competitive orientation (Interview 22). The other interview partners do not specifically mention any positive or negative experience or attitude towards cooperation with other governmental bodies. As the interview partners reveal very little information about interaction orientation towards other governmental bodies, I do not classify this into any category.

Another actor group mentioned by Carlos is *activists*. He states that he wants *activists* to consider his institution as an ally rather than an antagonist, although some *activists* have more radical thoughts then him. (Interview 22). However, since he declares himself as agreeing to their basic values, I classify his interaction orientation towards *activists* as cooperative ($U_x = X + Y$). The other interview partners do not mention any activist group in specific. Eric warns that one should not believe in the media and other propaganda. He claims that the Petorca society does not suffer from water scarcity as much as they say and defends large-scale agriculture companies. Thus, his orientation of interaction towards *activists* seems to be competitive ($U_x = X - Y$) insofar as their aims and solution strategies conflict.

The next actor group *large-scale agriculture* is mentioned in each interview. Carlos believes that *large-scale agriculture* overexploits water resources and states that he had conversations with representatives of large-scale agricultural companies. He tells that large-scale farmers claim to be »representative of the agricultural world« which in his point of view is misleading because they just represent a small but powerful part (Interview 22). Because he repeatedly expresses aversion towards *large-scale agriculture* that, according to him, hinders sustainable water management, I classify his interaction orientation as competitive ($U_x = X - Y$). Eric, on the other hand, calls overexploitation by large-scale agricultural companies a lie and strives for more investments of large-scale agricultural companies in Petorca Province (Interview 20). As he works for an institute that promotes economic development and tends to follow a win-maximizing ideology, the gains of agricultural companies simultaneously increase his gain. Hence, his interaction orientation towards *large-scale agriculture* is cooperative ($U_x = X + Y$). Eva mentions overexploitation by large-scale agricultural companies as a reason for the water scarcity (Interview 21). However, she does not reveal any hostile interaction orientation. Hence, there are not sufficient indicators about her orientation towards large-scale agriculture.

The remaining actor group *civil society* is mentioned mainly by Eric and Carlos. In specific, they refer to small-scale farmers and WUOs. Carlos repeatedly positions himself by defending the right to water of small-scale farmers and households and reports that his organization has a special department for counseling WUOs (Interview 22). Hence, I classify his interaction orientation towards *civil society* as cooperative ($U_x = X + Y$). Eric shows resentments towards small-scale farmers and *civil society* as he accuses them for being corrupt and blames them for exaggerating about the water scarcity (Interview 20). However, he also demands higher investment in training, education, and information, to improve cooperation between users. Although he does not agree with the concerns of small-scale farmers, he does not intend any loss for those users. Therefore, I will classify his interaction orientation towards small-scale farmers as selfish-rational ($U_x = X$).

7.5.3 Actor group: economy

The group *economy* consists of actors from the private economy. The biggest economic actors in the Petorca water conflict are agricultural companies and the water provider ESVAL (cf. chapter 7.4)[39].Therefore, this category is divided into two subgroups named *large-scale agriculture* and *water provider.*

7.5.3.1 Large-scale agriculture

The interviews of the category *large-scale agriculture* were held with two owners of large-scale agriculture companies in the Petorca Province: Víctor who owns land in the Petorca District, and Marco who owns land in the districts of La Ligua and Cabildo (cf. Table 16). According to my definition, large-scale agricultural companies own cultivated land that reaches over 70 ha.[40] However, it needs to be mentioned that the land cultivated by Víctor has decreased to less than 70 ha because of the water scarcity. The third interview analyzed in this category was conducted with a lawyer, Javier, hired by a large-scale agriculture association (LSAA) in Petorca. The two agricultural entrepreneurs, Víctor and Marc, are classified as aggregated because they do not form part of an organization that will be directly affected by their statements or actions. Although both belong to the LSAA, I will not categorize them as collective because the LSAA does not have hierarchical structures that imply direct consequences of their actions and statements for the other members.
Furthermore, especially Víctor does not identify with that group. Contrary to that, Javier's social unit is classified as cooperative because he gave the interview in his position as a lawyer contracted by the LSAA for which his actions should be independent of his personal benefits and serve the association.

---

39 As small-scale farmers make up a large part of society, they are represented in the group of civil society and not in the group of economy.

40 I use the same reference for big-scale farmers as the LSAA so that the interviewed persons coming from the LSAA would identify themselves as such, too.

Table 16: Large-scale agriculture. (Own research)

No.	Name (fictitious)	Sex	Age	District	Profession
23	Víctor	M	65-80	Cabildo	Large-Scale Farmer
24	Marc	M	50-64	La Ligua	Large-Scale Farmer
25	Javier	M	50-64	Lawyer	Lawyer

OVERVIEW

Marc and Víctor report to have lost cultivated land due to the water scarcity. Marc lost approximately one third of his plantations, and Víctor lost almost everything. All interview partners show their self-image as businessmen by putting economic progress in the center of their preferences. Further, they highlight feelings of both joy and sacrifices which, according to them, the life as an agricultural entrepreneur brings along. Marc and Javier are optimistic about the future of the Province. Víctor, in contrast, openly shows his frustration about the situation. According to him, not only climate change, but corrupt governmental bodies and the excessive expansion of large-scale agricultural companies have caused the ruin of the whole province. Marc and Javier basically refer to climate change as the main driver of the scarcity. However, Javier also blames the governmental bodies for being too weak to properly monitor water use in Petorca Province. All interview partners mistrust social movements in Petorca. They suspect that the activists hide their true motives. Therefore, the interaction orientation towards *activists* varies between hostile in the case of Javier, competitive in the case of Víctor and selfish-rational in the case of Marc. Further, Víctor and Javier claim to have cooperative orientation of interaction towards *civil society*. Marc, on the other hand, shows a selfish-rational interaction orientation. Since Víctor criticizes *governmental bodies* for causing water scarcity and shows a strongly negative perception, his orientation of interaction towards *governmental bodies and authorities* is classified as hostile. Javier and Marc show a rather selfish-rational interaction orientation towards that actor group. Besides, the interview with Marc was the shortest. He answered on point without revealing any further information. Javier and Víctor instead seemed to enjoy the conversations, and both invited to meet further on. Both claimed to be openly and direct on their thoughts about the situation in Petorca (Interviews 23-25).

MOTIVATIONAL ASPECTS

Both large-scale farmers are personally affected by the water scarcity. Marc explains that an area of more than 130 ha of planted fields has been lost completely. His remaining two fields which together reach approximately 400 ha, were saved by technical investments. Víctor does not give information about the entire area of the formerly cultivated land. However, he says that besides a field of citrus fruits, he lost more than 70 ha of avocado plantations. His biggest losses occurred around 2010 and 2015. After trying out different crops on experimental fields, he now cultivates green vegetables in greenhouses. While Víctor stresses to never have received subsidies by the state, Marc says that the last time his company was supported by a state's program was between 2005 and 2010 (Interviews 23, 24).

With regard to the actors' identity, the following information was revealed. Javier seems to have a strong connection with the Petorca Province. His family has a tradition of being politicians and Javier himself also ran up for being member of parliament. However, he was not successful. Proudly he confirms his dedication to the province by underlining that he – in contrast to other candidates - did not leave the province just because he did not win the elections. Furthermore, he claims to be good friends with different right-wing politicians and shows a tendency to prefer the right-wing government of Sebastian Piñera over the center-left government of Michele Bachelet (Interview 25). Marc and Víctor claim to support no political party or tendency (Interviews 23, 24).

Moreover, all three interview partners demonstrate strong identification with their business and their life as businessmen. When Marc is asked about his visions and values, he underlines that the main goal is to produce goods in an environmental and employer friendly manner, while he is always searching for new species that develop faster and achieve higher profitability (Interview 24). Similarly, Víctor states to be an environmentalist, which he implements in his work by not using certain pesticides and fulfilling all norms for the foreign supermarkets. Furthermore, he claims to care for his workers (Interview 23). The lawyer Javier comes from a family of farmers and believes that agriculture is and should be an increasing driver of the Chilean economy. He sees himself and his clients as persons who bring progress to Chile and underlines that his effort should be appreciated by politics and society (Interview 25). All of them are proud of their work and claim to enjoy it. At the same time, they consider it as complicated and sacrificing. Despite their main interest in economic growth, they cultivate their self-image as benefactors to the society. They highlight that they not only bring progress and job opportunities, Víctor and Javier also say that they offered help to the municipality and to the government respectively. However, their help was not accepted (Interviews 23-25).

The emotional perception of the situation varies among the interview partners. Javier underlines his emotional connection with the agriculture business and with the persons behind it. His voice changes to underline his feelings when he talks about the economic loss of farmers in Petorca Province. Nevertheless, he repeatedly declares to be optimistic and believes that suitable solutions will be found (Interview 25). Likewise, Marc has an optimistic view and hopes to increase his business (Interview 24). In contrast to them, Víctor shows his frustration. He uses harsh words for describing the situation and expresses his anger about those who he thinks are responsible for the scarcity. Moreover, he claims that he has the reputation of being the »crazy water man« because he warned about the water scarcity in early times (Interview 23).

COGNITIVE ASPECTS

All interview partners complain about the same water scarcity impacts: economic damage of large- and small-scale farmers because of harvest loss. Furthermore, they observe a rise of unemployment and decreasing population due to emigration into other regions. Marc also mentions increasing labor costs because of this emigration (Interviews 23-25).

All interview partners believe that climate change is the main reason for the water scarcity (ibid.). However, Marc underlines that he does not believe that agriculture affects climate change (Interview 24).[41] Javier and Víctor agree that agriculture does influence the water scarcity in Petorca. Both stress that the area has been overplanted and that the water resources are overexploited (Interviews 23, 25). Nevertheless, their perception of this situation varies. Rather than mismanagement, Javier believes that this is an ordinary process insofar as the rules of nature and the rules of market will always lead to equilibrium. Several times he mentions the importance of environmental rules and simultaneously highlights the importance of market rules (Interview 25). For example, to the question of what could be a key to change of the current situation, he answers:

---

41   Although at least 14 % of global greenhouse emission emerge from agriculture
      (OECD Observer 2010).

»Es que el cambio lo va a decir el mercado. No vas a volver a plantar en lugares donde cada cierto tres cada cuatro, cinco años va a haber una sequía y vas a perder tu plantación. Ningún inversionista ningún agricultor está dispuesto para perder dinero. ¿No cierto? Entonces vamos a volver naturalmente.« (Interview 25)

(»The market will dictate the change. It puts us back into place where every three or four or five years there will be a drought, and you will lose your plantations. No investor, no farmer is willing to lose money, right? Hence, we will return naturally.« Translated by author)

This argumentation indicates that he perceives economics as a natural science which rules are imposed on society like physical laws and that need to be respected in order to achieve well-being. His views seem to be close to the neoliberal ideas of a spontaneous order (cf. Hayek 1999). Víctor, on the other hand, expresses anger and frustration about the overexploitation of water resources by large-scale agricultural companies (Interview 23). In his eyes, corruption and the lack of monitoring and sanctioning of agricultural activities has led to an economic behavior that disrespects nature's limits. As he says:

»Y ahí me dio cuenta de que estos seguirán creciendo. Seguirán grandes agricultores haciendo plantaciones gigantescas. Y todo el mundo, los pequeños agricultores, plantando. Claro, la palta es un buen negocio y que sé yo. ¿Cómo se puede seguir plantando? ¿Cómo están capaces de hacer inversiones gigantes y estar sacando agua donde no tienen derecho?« (Interview 23)

(»And then I realized that they will keep on growing. Big farmers will keep on doing gigantic plantations. And everybody, small farmers, planting. Of course, avocado is a good business and whatever. How can they keep on planting? How are they able to make these gigantic investments and drain water where they do not have the rights?« Translated by author)

His understanding of economics and nature is different from the idea of Javier insofar as Víctor does not perceive the overexploitation as a natural process. Consequently, their perceptions of the current water regulation and institutional processes also vary. Javier does not think that the socio-economic situation of a person influences on his or her access to water. He believes that Chile is a country of water abundance and that the main problem consists of the unequal geographical distribution of water. He likes the current water regulation and believes that especially water privatization and trade brought progress to the country and will balance the production of agricultural goods according to the available water resources (Interview 25). Javier argues like that:

»Es cada vez más escaso. O sea, el agua va a tener cada vez mejor, más precio, ¿no cierto? Y por lo tanto, no van a haber producciones agrícolas que puedan financiar el valor real del agua.« (Interview 25)

(»It's increasingly scarce. Well, water will have an increasingly better, higher price, right? Consequently, there will be no agriculture production that could finance this real value of water:« Translated by author)

Although he concedes that everyone should have access to drinking water, he states that water property is and should be exclusive. Despite of this overall positive opinion on the current water management, he believes that more effective fees should be charged of private users who block the water market by not using their water rights.[42] Furthermore, he criticizes the DGA for being financially weak and unable to pursue cases of irregular water use. His main criticism is about the government. He believes that especially the government of Bachelet blocked agriculture progress, for example by blindly declaring rivers as dried out without sufficient evidence. In his eyes, the government should invest more in water infrastructure and other needs of the agricultural sector, while at the same time respect the free market and not intervene entrepreneurs' management. Also, he believes that the lack of trust of society in governmental bodies and personal or political interests occasionally blocks economic progress. For example, according to him, the temporarily introduced law which allowed small-scale farmers to legalize their wells, informally called »ley de mono«, was not efficient because small-scale farmers did not trust the process (Interview 25).[43]

Also, Marc is satisfied with the current water regulation and governmental bodies. He does not believe that the regulative framework has a negative impact on the water scarcity. He declares that his company never confronted any problems, neither with the regulation nor with governmental bodies. However, he and Víctor believe that the socio-economic status of a person influences his or her water access, for example, it affects the ability to technically adapt to the water scarcity (Interview 23, 24).

Nevertheless, Víctor's conclusions go further. He believes that the lack of regulation and corruption have led to the overexploitation of water. He is not content with current water governance but criticizes the misman-

---

42 A fee for non-water-use was introduced in 2005 and is criticized in the general debate for being too low. Discussion of raising the fee emerged in the debate of the new water code reform (cf. chapter 7.4).

43 For more information on *Ley de Mono* please read chapter 8.2

agement of the DGA. He reports about cases of conflict of interests and corruption that has led to oversharing of water rights and to illegal extraction of water resources (Interview 23). To illustrate his point of view, he tells the interviewer about his personal experience with a legal expert who tried to bribe him:

> »Un día llegó un señor a mi casa, el perito judicial, aquí mismo. Estuvo sentado ahí. Y me dice 'yo soy perito judicial. [...]. Yo sé que tú tienes un campo enorme, me dice. Te vengo a proponer inscribirles derechos de agua.' '¿Cómo?' le digo yo. 'Sí', me dice. 'Pero no te asustes, tú me lo respaldas una vez que están inscritos. [...], hacemos un pozo un pozo que tu tengas y yo certifico que son antiguos, presentarlo antes del juez, todo el trámite y toda la cosa, y el juez se los va a entregar si yo le certifico el pozo antiguo y te da 30 l/s.' « (Interview 23)

> »One day a man came to my house, a legal expert, right here. He sat down over here. And says: 'I am a legal expert. [...] I know that you have a big field', he tells me. 'I came to suggest registering water rights.' 'What?' I said. 'Yes', he says. 'But don't get scared, once they are registered you pay me off. [...] we will take a well, a well that you have, and I will certificate that its old, present it to the judge, the whole paperwork and everything, the judge will give you [the water rights] if I certify that it's an old well and he gives you 30 liters per second.' « Translated by author)

Moreover, he believes that the government was wrong when it implemented the »*ley de mono*«. According to Víctor this law was using the name of the small-scale farmers (in his words »poor«) to legalize wells of large-scale agriculture companies. Consequently, it worsened the water scarcity (Interview 23). As he reports:

> »Un solo usuario, uno, un usuario pidió 90 pozos. 90 pozos ilegales, uno. 90 por 2 litros son 180 litros. [...] Eso era un usuario. Imagínate la cantidad de usuarios que pudieron. El río quedó absolutamente sobreexplotado. [...] Entonces yo ahí le digo »Mira. En esta vía en el nombre de los pobres se ha cometido la peor anticipada del mundo. Por beneficiar a los pobres al final terminas cagando a todos. Y eso es lo que va a pasar aquí.« Todos los pobres de ese valle murieron con esta ley.« (Interview 23)

> (»Just one user, one, one user asked for 90 wells. 90 illegal wells, one. 90 times 2 liters are 180 liters. [...] This was one user. Imagine the number of users that have been able to do that. The river remained absolutely overexploited. [...] So, I said. »Look, this was the worldwide worst anticipated road to take in the name of the poor. To benefit the poor, you end up blowing it for everyone. And that's what is going to happen here.« All the poor people in this valley died because of this law.« Translated by author.)

According to Víctor, the biggest problem concerning water regulation is the lack of compliance wherefore he has no planning security. He claims to have no trust into the system under the current conditions of corruptive management and low monitoring of water use. Nevertheless, he does not criticize privatization of water resources. He fears that an alternative regulation, such as water nationalization, could give power into the hand of corrupt politicians and lead to a totalitarian state similar to the government of the GDR (Interview 23).

Consequently, Víctor shows mistrust against the *activists* as he strongly associates the idea of nationalization of water resources with them. He expresses his fear about *activists* taking away his water rights, and declares to be an honest man who never extracted water illegally (Interview 23). According to him, the idea of nationalizing water is not realistic, he expresses his frustration about people who pretend to provide easy solutions and complains about opportunistic behavior. He says:

> »Pero veo que estamos llenos de vendedores de la pomada acá en Chile. ¿Sabes que es la pomada en chile? Es como que te da soluciones que son puros palabras. Y aquí me tocó ver, dar peleas y se murieron todos. Y porque nadie hizo nada. Y uno cuando trataba hacerlo, te dabas cuenta de que el otro tenía otro interés.« (Interview 23)

> (»But I see that Chile is full of ointment sellers [quacks]. You know what ointment is? Is when you give solutions that are only words. And here I have witnessed and fought, and everyone died. Because no one did anything. And when you tried to do something, you realized that the other person had another interest behind.« Translated by author.)

Javier and Marc share this fear of the *activist's* hidden political or personal interests. Marc suspects that they are underinformed and probably seek for a political career. Javier refers directly to MODATIMA when talking about *activists* in Petorca Province. He also thinks that their goals are political. Javier believes that members of that group lie and take advantage of small-scale farmers by making them believe that an alternative water management system is possible; however, according to Javier, the *activists* pursue only their own personal career. For instance, he says that they want to expropriate water owners but do not openly say that this expropriation also includes small-scale farmers: »*Porque MODATIMA les dijo que el agua era de todos. Y no es así.« (Interview 25) (»Because MODATIMA told them that water would belong to everyone. And it is not like that.«* Translated by the author). Further, he thinks that the group's main argument is the blaming of large-scale agriculture for overexploitation of water by illegal extractions. Therefore, it's members strategically accuse politically significant persons of water robbery without any evidence in order to stir up hatred against financially and politically powerful persons. According to Javier, the number of illegal water extractions in Petorca Province is insignificantly small. Moreover, he stresses the *activists'* incredibility and reports that members of MODATIMA themselves have been accused of water robbery. According to him, this is another proof of their false strategy. At the same time, he thinks that this movement is too weak to succeed with its goals because, firstly, their ideas have no political support in Chile and, secondly, he thinks that its number of followers is decreasing because of the rain that alleviates the conflictive scarcity (Interview 25). Moreover, he believes that the *activists* do not know how to create welfare and should leave his actor group namely large-scale companies in charge of welfare production:

> »Y esto está diciendo gente que nunca produjo riqueza. No saben producir riqueza. Y están cortando la libertad de la gente que si sabe producir riqueza. Déjenlos libres. No nos molesten demasiado. Si cumplen las leyes *ciao*, déjenlos tranquilos. (Interview 24)

> (»And these people who are speaking never produced wealth. They don't know how to produce wealth. And they cut the freedom of people who do know how to produce wealth. Let us be free. Do not annoy us too much. If the laws are complimented *ciao*, leave is alone.« Translated by author.)

With regard to conflict solutions, the interview partners have different ideas. While all of them mention technologic innovations, only Víctor remains skeptical. According to him, these technologies will not help if a well working judicative and executive system does not exist. He repeatedly stresses the importance of transparency, law compliance and sanctions (Interview 23). As indicated above, Javier focuses on technical solutions. In his eyes, public-private partnerships and better management of the global food market will lead to a progressive future of Chile. He believes that a *water highway* should be built in the future. That technology will bring water from the south to the north of the country.[44] In this way, he believes, the south of Chile will pay back the wealth that is created by the north. Further, he believes that technical solutions, such as water tanks, should be provided to small and medium scale farmers and demands a second *ley de mono* to give small-scale farmers the possibility to legalize their wells as in his eyes they own most of the illegal wells (Interview 25). Marc says that more political will is needed to install technical solutions. He criticizes the last governments for not having invested enough money into the Province and suggests the installation of water tanks, water roads and desalinization plants to increase the water supply (Interview 24).

---

44   For more information about the water highway please read chapter 7.1

ORIENTATION OF INTERACTION

The interview partners mention four other actor groups: *civil society*, referring to *activists* and *citizens, governmental bodies and authorities* and *large-scale agriculture*. Only Javier reports to have contact to all these actor groups including to the *activists*. He says that he discussed the water scarcity with members of MODATIMA and does neither agree with their perception of its causality nor with their solution strategy. He accuses the group of intentional slurring on the entrepreneurs' reputation, of stirring up hatred and of taking advantage of the impoverishment of small-scale farmers (Interview 25). One of his dominant aims revealed in the interview, is to prove the *activists* wrong. Consequently, what is profit for them would mean loss to Javier ($U_x = -Y$). Considering their contrasting aims, as he is the lawyer of large-scale agricultural companies, his interaction orientation towards *activists* is classified hostile. Víctor has similar doubts about the true intentions of the *activists* as Javier. Although, he does not express such negative perception nor accuses them of harming society intentionally, he still fears that the aims of the activists could harm his business and liberty (Interview 23). Because he does not show a hostile orientation but contrasting aims and fears towards the *activists*, I classify Víctor's interaction orientation as competitive ($U_x = X-Y$). Marc emphasizes the need to keep in mind that these conflicts mostly evolve from different world views. He does not express any anger or fear but does not share the views of the *activists* (Interview 24). Due to his rational answers that do not show any contrasting nor overlapping interests with the *activists*, I classify his interaction orientation as selfish-rational ($U_x = X$).

The next actor group is *citizens*, respectively small-scale farmers. Javier reports that via the LSAA there have been various common projects and meetings with small- and medium scale farmers. Víctor says that he cooperated with his neighbors several times to help with water infrastructure or to report water robbery. As demonstrated in the motivational aspects, both Javier and Víctor, underline that society's welfare depends on the large-scale companies' activities. Additionally, they perceive of themselves as superior benefactors. Hence, rather than seeing societies gain as their own, they see their own gain as benefitting society. In that sense, they do not aim at the society's gain but see it as a positive side effect (Interview 23, 25). Considering that, their interaction orientation towards civil society and small-scale farmers is classified as selfish-rational ($U_x = X$). Marc only rarely refers to small-scale agriculture or civil society. He basically considers society as a potential workforce and claims to provide decent jobs. Besides, he does not reveal any cooperative nor hostile behavior.

Therefore, I also classify his orientation of interaction as selfish-rational ($U_x = X$).

The interaction orientation towards *governmental bodies and authorities* varies between the interviewed persons. Since Marc claims to be content with the *governmental bodies* and does not seem to have any further positive or negative ambitions, I again classify his interaction orientation as selfish-rational ($U_x = X$). Javier reports about several meetings between the LSAA and different public water institutions which he believes helped to slowly progress in installing different projects and to advocate the interests of large-scale agricutlure (Interview 25). However, he believes that the private sector should have more freedom of development and complains about several restrictions especially during the government of Bachelet. Therefore, the interests of the *governmental bodies* do not necessarily overlap with the aims of the LSAA and Javier and his interaction orientation is rather selfish-rational ($U_x = X$). Víctor, in contrast, confronted representatives of *governmental bodies and authorities* several times. He repeatedly accuses them of being corrupt and accomplices of the water overexploitation in Petorca Province. Furthermore, he points to the mismanagement of the DGA and claims to have no trust in public authority or in the government. He underlines his independence from the state by stressing that he never received subsidies. However, he claims to have tried to cooperate with the local government by offering technical solutions but has been denied by the municipality (Interview 23). Those experiences have led him to a negative perception of *governmental bodies* on the local and on the national level as the following statements demonstrate:

>»Lo que yo me di cuenta de que eso es una estafa a nivel nacional. Que es un robo institucionalizado.« (Interview 23)

>(»I realized that it's a scam on a national level. It is institutionalized robbery.« Translated by author.)

>Pero sí, fue un weón por haber plantado tanto y por no haberme dado cuenta que esto es un país razonaría, egoísta, y con unos senadores y diputados que son una vergüenza absoluta. Y con directores de agua y organizaciones institucionales de agua que son...Que no sirven para nada. No existe control. (Interview 23)

>(»But yes, I have been an idiot planting that much and not realizing that this is an egoistic country, with senators and congressman who are a complete embarrassment. And with water managers and water institutions that are useless. There is no control.« Translated by author)

With regard to Víctor's accusations and his anger towards public institutions, I classify his orientation of interaction as hostile ($U_x = - Y$).

The last actor group is *large-scale agriculture*. Marc declares to have a friendly relationship without any rivalry towards his business fellows and that communication via the LSAA helps to cope with the water scarcity (Interview 24). Being the lawyer of the asscociation, the interests of Javier are in line with the interest of large-scale agricultural companies. During the interview he highlights the welfare that is produced by those companies and defends the companies' liberty to act in a free market. Therefore, I will classify Javier's and Marc's interaction orientation to large-scale agriculture as cooperative ($U_x = X + Y$). Although Víctor claims to have no problem with anyone, his image of large-scale agricultural entrepreneurs seems to be hostile. He reports about conversations with an entrepreneur in which he blamed him and his fellows for buying the Chilean justice system and for ruining everyone's chances to cultivate in the valley. Moreover, he explains to have reported illegal water extraction by large-scale agricultural companies several times. Víctor took part in meetings of the LSAA. However, he does not have a good impression of these meetings, as he believes the members are arguing over water (Interview 23). In Víctor's view, the behavior of large agricultural companies affects his interests, and I will classify his interaction orientation between competitive and hostile ($U_x = -Y / U_x = X - Y$).

7.5.3.2 Water provider

The water provider in charge in Petorca Province is ESVAL - a private company which main shares belong to the Canadian pension plan for teachers called OTPPB. ESVAL provides water for the urban area while the rural area is covered by community water user organizations (APR). The interview was held with Óliver (cf. Table 17). He works in the department of public relations in the ESVAL head office in Valparaíso, hence is social unit is classified as cooperative. The interview took place in the waiting hall and Óliver preferred not to be recorded by audio tape. The interview was rather short (approximately 15 minutes).

Table 17: Water provider. (Own research)

No.	Name (fictitious)	Sex	Age	Profession / Position
26	Óliver	M	50-64	Employee at ESVAL

MOTIVATIONAL AND COGNITIVE ASPECTS

Óliver does not express any personal motivational aspects and limits his statements to representing the company. For example, he does not give his opinion on the current debate about the water code reform and indicates that ESVAL does not position itself in the discussions. Nevertheless, he states that water privatization has brought progress to Chile insofar as the supplied area has increased owing to higher investments since the water privatization began. Furthermore, Óliver states that, with regard to drinking water provided by ESVAL, there is no unequal access due to socio-economic status of the user. He explains that ESVAL covers 98% of the households belonging to the company's area of responsibility and provides subsidies for users who struggle with paying the water bills.

According to Óliver, water scarcity in Petorca Province is caused by climate change and increased because of an augmented consumption due to demographic growth. Further, he criticizes that there is too little control on irrigation water. He complains that the lack of formal community organizations challenges cooperation with civil society. Following these observations Óliver concludes that a solution to the current conflict is an improved water user organization in the form of a Junta de Vigilancia (Óliver 33)[45]. Consequently, he underlines the importance of civil society's participation and awareness in order to save water. Moreover, he demands technical innovation. However, he believes that a desalinization plant should be the last option as the inversion is too cost-intensive.

ORIENTATION OF INTERACTION
With regard to other actor groups, he mentions *large-scale companies* (not specified), *governmental bodies and authorities, civil society* and *activists*. As the interview was rather short, assumptions on the orientation of interaction can only be made with reservations that there may be undiscovered information. Óliver refers only once to large-scale companies when he is asked about the main challenges of ESVAL. He responds that ESVAL must negotiate with other companies that aim at buying water rights. This statement indicates a competitive orientation of interaction ($U_x = X - Y$). Further, Óliver reports about his positive experience in cooperating with *governmental bodies* and common projects. Hence, the orientation of interaction can be classified as cooperative ($U_x = X + Y$). In several parts of the interview, Óliver refers to campaign strategies that address the *civil society*. However, the interaction seems to be rather top-down than bottom-up as he only mentions information campaigns but not active participation strategies. Moreover, civil society is the main client of ESVAL,

---

45   More information about Junta de Vigilancia to be found in chapter 7.4.

which makes both actors dependent on each other while following opposite aims: On the one hand, ESVAL provides water and benefits from high water costs; the users, on the other hand, need water provision and benefit from low water costs. Therefore, I classify the interaction orientation from ESVAL towards water users as selfish-rational ($U_x = X$). The last actor group mentioned is *activists*. Óliver declares that ESVAL has tried to initiate conversations with them in working tables. He mentions the improvement of monitoring and sanctioning of water use as a common interest. However, he calls the relationship »difficult« because of the opposing perception of the water privatization. Since the opposing aims seems to dominate the relationship, I will classify the interaction orientation as competitive ($U_x = X - Y$).

7.5.4 Others

This subchapter demonstrates the action orientation of representatives from different institutions which could not be categorized into public, private or civil society. Thus, each expert will be analyzed separately. These interviews were of special interest to this study because the interviewees are not directly affected by the water scarcity, but they are considered as special experts. They have been working on the conflict from a meta level while taking part in different actions and gaining different perspectives. The importance of each interview partner's perception to the case study will be explained in the subchapters.

7.5.4.1 Padre Fernando

For approximately 10 to 15 years, Padre Fernando has worked as a catholic priest in the Petorca District (cf. Table 18). As he functions as a representative of the catholic institution, his social unit is classified as cooperative. The interview with Padre Fernando was of special interest for several reasons. Firstly, due to his work as a priest and Chile being a catholic country, he is supposed to have a wide range of influence. Secondly, he appeared in various documentaries about the conflict. Thirdly, pope Francisco stated in the *Enzycilca* of 2015 that water should not be privatized and that water access needs to be respected as a human right (Franziskus 2015). Therefore, it was of additional interest to know in how far this statement of the international church leader influences the actors in the case of Petorca.

Table 18: Padre Fernando. (Own research)

No.	Name (fictitious)	Sex	Age	District	Profession / Position
27	Fernando	M	65-80	Petorca	Priest

MOTIVATIONAL AND COGNITIVE ASPECTS

The motivational aspects of Padre Fernando are characterized by his identification with the role of an education provider and his interest in spreading values of the catholic church. He stresses that the catholic church helps people in Petorca primarily by educating them about their rights and underlines the importance of equality by saying that he does not like to distinguish between rich and poor. Padre Fernando recognizes that most church representatives belong to the socialist party PS (Partido Socialista). However, he states to belong to no political party. He struggles with both sides, the political left, and the political right.

Padre Fernando mentions three reasons for the ongoing water scarcity: drought, climate change and expansion of avocado plantations. He explains that droughts are a typical phenomenon in the area which recurs cyclically every 5 to 10 years. However, he underlines that expansion of large-scale agriculture and the introduction of the highly water consuming fruit avocado, have worsened the scarcity. He accuses large-scale agricultural companies of overexploiting the resource by planting on hills as well as by building illegal drains and wells. According to him, the regulative system does not provide equality to its users. He claims that the owners of those agricultural enterprises are congress members and instead of representing the people, they would persue their personal economic benefits: »*Ahora nos dimos cuenta de que ellos gobiernan para ellos, no para la gente*« *(Interview 27). (»Now we realized that they are governing for themselves, not for the people.*« *Translated by author.)*

He concludes that although small-scale farmers are well organized and supported by governmental bodies, they do not obtain financial power to prosecute water. Padre Fernando describes that small-scale agriculture is decreasing and more and more people are selling their water rights and migrating to the north. He reports that especially men leave in order to find work in the mining industry. Those men often form new families causing financial and social damage to the women and children left in Petorca. Further, Padre Fernando does not believe that the conflict about water scarcity will alleviate in the future. Instead, he believes that conflicts over water and other natural resources will rise.

He underlines that education is the key to changing this situation and states that people need to be educated in order to vote for better representatives. Only then, he believes, will the government introduce technical solutions such as dams to provide water and turn water into a common good that is owned by the community. However, he does not believe that the church could have an impact on this development as he repeatedly describes its negative influence on society. He stresses that an increasing number of citizens has converted to protestant churches and that the pope has lost influence on local communities (Interview 27).

ORIENTATION OF INTERACTION

Padre Fernando refers to different actor groups during the interview: *activists, civil society, large-scale agricultural companies* and *governmental authorities*. Mostly, he mentions *civil society,* which in the Petorca District consists mainly of small-scale farmers. While stressing his sympathy for that group, he explains that small-scale farmers have their own closed culture and organization, because of which he does not have a strong connection to them. Nevertheless, he devotes care to their traditions and the social wellbeing of the community. Accordingly, I classify his orientation of interaction towards *civil society* as cooperative ($U_x = X + Y$).

Despite of some cooperation with governmental bodies, Padre Fernando reveals to have a negative image of governmental authorities. As already mentioned, he regards them as complicit to the water scarcity because of the lack of regulation and conflicts of interests. Hence, his orientation of interaction towards *governmental authorities* varies between selfish-rational ($U_x = X$) and competitive ($U_x = X - Y$) depending on the institution or representative.

Towards *large-scale agricultural companies* Padre Fernando does not seem to have any active interaction. Although, he believes that the expansion of those companies worsens the water scarcity, he expresses more anger about politicians who allow this situation to happen than about the companies. However, Padre Fernando's pursuit of the society's wellbeing seems to be jeopardized by large-scale agriculture. Therefore, I classify his orientation of interaction towards agricultural companies as competitive ($U_x = X - Y$).

The last actor group mentioned is *activists.* Padre Fernando says that he has been interviewed by representatives of social movements before. However, he does not answer in how far they cooperate. He appears to agree with their goals but seems to be pessimistic about the chances of their success. He believes that the power of private industries, which work together with governmental representatives, is too big to be challenged. Due to their common goals and the occasional cooperation, I classify Padre Fernando's interaction orientation towards activists as cooperative ($U_x = X + Y$).

7.5.4.2 Susana

Susana is a Chilean politician and activist working in the area of environmental protection for more than 25 years. By the time of the interview she is executive director of a non-governmental, non-profit organization aiming at promoting environmental sustainability in the legal and institutional framework of Chile (cf. Table 19). The organization focuses primarily on water and energy issues. It counsels politicians and provides educational workshops to communities and cooperates with social movements nationwide. Since Susana represents the NGO, I classify her social unit as cooperative. Hence, her actions are supposed to be independent of her own benefits. The view of Susana on the conflict is interesting for this study because she is experienced in working on environmental topics in Chile from a macro-level and reflects the perception and orientation of an influential and well-known NGO in Chile. Furthermore, the NGO has been working on the general topic of water conflicts for more than a decade, for which Susana may provide a more time-independent view.

Table 19: Susana. (Own research)

No.	Name (fictitious)	Sex	Age	Profession
28	Susana	F	65-80	NGO Director

MOTIVATIONAL AND COGNITIVE ASPECTS
Susana's identity is characterized by her work as an environmentalist. During the interview she underlines the importance of social and ecological sustainable water management several times. According to her, the water conflicts that Chile faces are generally caused by the competitive demands of ecosystems, citizens and extractive industries, such as mining, agriculture-export or hydro energy. Susana is convinced that the ground of this situation is set by the state's policy that started during the dictatorship of Agosto Pinchot and since then has focused on liberal markets and diminishing the state's power. She criticizes the water privatization and the nonexistence of priority of water for domestic use and ecological safety. Moreover, she believes that economic players have too much influence in the congress. Specifically in Petorca Province, she claims that water gov-

ernance has been undermined by a corrupt minister who at the same time owns an agriculture industry in the province. As she says:

> »[…] estaba Pérez Yoma ahí, y todos miraban para el techo, si tú te enfrentabas a Pérez Yoma que era el ministro del interior, te ibas para tu casa, perdías el trabajo, y el parlamentario ahí cerca demócrata cristiano era parte de los que robaban agua.« (Interview 28)

> (»[…] there was Pérez Yoma, and everyone looked away, if you would have confronted Pérez Yoma who was the minister of the interior, you would have been fired, you would have lost your job, and the parliamentarian close to the Christian Democrats was one of those who stole water.« Translated by author)

Moreover, she reports that the state did not comply with the law in Petorca Province as the water rights were overshared by the DGA.

According to Susana, the solution for these conflicts is a change in public policies. She believes that, due to the lobby of economic players, a rapid change towards sustainability and equal water access is not achievable. Therefore, she stresses an incremental change by first demanding law compliance, then a water code reform and thirdly a change of the constitution. Therefore, her NGO counsels the congress's Commission of Water Resources and various politicians. In addition, she believes that a rise of awareness in society is needed to achieve such a change. She says that her organization strives to educate communities about their water rights and to unite different environmental and social movements to express one common demand. She believes that both, policy and society, do not pay attention to needs of people living in rural areas and that environmental activism is needed to raise that lack of awareness. She argues: »*Si no hay movilizaciones el estado no hace nada.« (Interview 28). (»If there are no movements, the state doesn't do anything.« Translated by author.)* In addition, she explains that weather conditions critically influence the discourse as especially the winters rainfalls take away pressure to change the law because people tend to believe that climate change has stopped.

ORIENTATION OF INTERACTION

During the interview Susana mentions several different actor groups. With the idea of joining power and strengthening demands, her NGO cooperates closely with different communities and social movements from all over Chile in the form of publications, workshops and other projects. Regarding Petorca Province, Susana claims to be good friends with MODATIMA. Nevertheless, she believes that the group is working in a destructive manner because they mainly criticize and report water robbery and neglect to make positive proposals. Since Susana stresses that she wants to build common ground to create a lobby for the different community needs, the interaction orientation towards *activists* and *civil society* is classified as cooperative ($U_x = X + Y$).

Another actor group mentioned by her is politicians. Susana states that she counsels the Commission of Water Resources (*Comisión de Recursos Hídricos*) and other politicians. However, she also reveals to have had a bad impression of water management politics and criticizes the impact of lobbyism by economic players on politics. Despite this criticism, rather than organizing or acting against politicians, she tries to convince them to implement socially and ecologically sustainable changes in the water regulation. Hence, her goals depend on her cooperation with decision makers. Therefore, her orientation of interaction towards politicians depends on the individual politician's position and if it overlaps with her ideas or her ability to influence them. Hence, I classify her orientation of interaction as selfish-rational ($U_x = X$).

7.5.4.3 Maria

Maria is an agronomist specialized on counselling small-scale farmers (cf. Table 20). Marias interview is of special interest for this study because she has gathered knowledge about Petorca on a meta level while having been working in the Province for approximately 5 to 10 years. She is originally from another town outside of the Vth region. Maria has worked with many WUOs in Petorca Province. She overviews the social and working structures of different actor groups and is in contact with different governmental bodies. Hence, her expertise is characterized from an academic point of view as that of a meta level. Further, Maria can be classified as a cooperative actor because she works in a formally structured academic organization and her actions are supposed to be independent from her personal wellbeing.

Table 20: Maria. (Own research)

No.	Name (fictitious)	Sex	Age	District	Profession
29	Maria	F	20-35	La Ligua	Agronomist

MOTIVATIONAL AND COGNITIVE ASPECTS

Maria does not directly mention her values or interests during the interview. Nevertheless, her statements reveal that social equality is of primarily importance to her. For example, she stresses that a street worker is not of any less value than an academic. Further, Maria claims to be politically independent insofar as she does not feel represented by any existing party. Moreover, she shows a critical attitude as she underlines the importance of verifying information by personal research and expresses her mistrust in the media. At the same time, she is careful about making statements on topics of which she does not feel to have enough expertise or evidence.

Maria mentions two main reason for the water scarcity. Firstly, she refers to the geographical location of Petorca Province (no natural water from melting snow). She explains that those droughts occur every couple of years and is unsure about the influence of the anthropogenic climate change. Secondly, she mentions water overexploitation by large-scale agriculture that drains the groundwater aquifers. She believes that water access depends on socio-economic resources and explains that this situation roots in the separation of water and land rights by the current water code that favors economic powerful actors.

Maria states that the water scarcity aggravates the loss of small-scale agriculture and emigration as primarily young people leave to find work in other areas. According to her, this emigration has a negative impact on the WUOs. Within those organizations, she recognizes a lack of motivated members that take over leading roles. Instead, those organizations must cope with participants of low educational background and a high rate of analphabetism. She believes that there is little interest in cooperation because, firstly, WUO members feel powerless towards large-scale agricultural companies and, secondly, they are used to individual solutions and do not see the benefits of cooperation. To describe their orientation of interaction she repeatedly uses the term »individualist«. Furthermore, she states that the inhabitants of the Petorca District are more organized than those of the La Ligua District because. She thinks that people in the Petorca District are more hopeful because the scarcity is not as bad as in the La Ligua District and therefore, the inhabitants cooperate more.

Besides the lack of motivation to cooperate with other WUO members, Maria also reports about the mistrust towards governmental bodies. She says that high bureaucratical standards cause refusal of governmental programs. Moreover, she believes that discrimination of small-scale farmers by municipal bodies may have led to the lack of willingness to cooperate. Like the WUOs, Maria also criticizes governmental bodies for not cooperating. Again, she describes the behaviour of governmental bodies as »individualistic. Further, she criticizes the lack of monitoring and the low sanctioning of illegal water extraction. She states that sanctions on water robbery are so low that companies just include them into their business calculations. Additionally, she explains that governmental bodies are geographically distant from their clients, which is why, for example, small-scale farmers often do not have the time or the money to approach those institutions and to report water robbery. In addition, she claims that people mistrust any governmental, respectively political, authority because they suspect conflicts of interest or corruption. Maria believes that although the core problems are to be found in the water code, corruption and conflict of interest worsen the situation on each political level. In an informal meeting outside this interview she reports to have witnessed corruption at the local municipality herself.

This rejection of politics and of any political activity, she believes, causes rejection of social movements. Maria finds those social movements important to represent the needs of the citizens. Nevertheless, she claims that members of such movements pursue political interests. The following statement illustrates her perception:

> »O sea, yo siento eso. Yo creo que a mucha gente también le pasa. Cuando algo ya se politiza se genera ese rechazo. Mientras tanto, mientras no sea algo político y que en realidad si se defiende tu interés y se quiere luchar por cambiar algo, se apoye.« (Maria 62).

> (»Well, I feel that. I believe that many people perceive the same. If something is getting political, it produces a rejection. While it is not political, and they defend your interest for real, and they really want to fight to change something, it will be supported.« Translated by author)

According to Maria, improved user organization combined with improved technology could lead to a conflict solution. In addition, she believes that education is a key to alleviate the current and future conflicts. Maria assumes that most problems originate from individualistic behavior. She says that underlying norms and values for such behavior are taught by the society and in school, for which a change towards new values that highlight empowerment and equality will need to be introduced as a long-term aim.

ORIENTATION OF INTERACTION

Maria works closely with the civil society and specifically refers to small-scale farmers and other WUOs. She has built up close relationships and long-lasting friendships. Since she is interested in that actor groups well-being, her interaction orientation can be classified as cooperative ($U_x = X + Y$). Concerning large-scale agricultural companies, she reports to have no strong connection and that they would not be interested in cooperating. As she describes a rather distant relationship towards this actor group and neither reveals positive nor negative emotions, I classify her interaction orientation towards *large-scale agricultural* companies as selfish-rational ($U_x = X$).

Her orientation towards *activists* seems to be indifferent. Although she supports activism as such, Maria remains skeptical and does not show any cooperative nor hostile orientation respective actions. Therefore, I classify her action orientation again as selfish-rational ($U_x = X$).

Maria also talks about *governmental bodies and authorities*. Her statements endorse mistrust towards this actor group. However, Maria does not seem to be interested in any gain nor loss of that actor group and reports to be aiming at more cooperation and communication between *governmental bodies*. Thus, her interaction orientation towards governmental bodies and authorities can be classified as between selfish-rational ($U_x = X$) and cooperative ($U_x = X + Y$).

7.5.4.4 Alfredo

Alfredo is the owner of a retail chain in the Petorca Province (cf. Table 21). He moved to Petorca approximately 30 years ago and had a leading position in the large-scale farmers' association (LSAA) and in *Water Round Table* installed approximately in 2008 to mediate between different water users.[46] His point of view is interesting for this study, as he has been working on conflict resolutions as an interlocutor between different actors. Additionally, due to his former position in the LSAA, he is especially experienced at working with representatives of medium and big scale agricultural companies. In this study, he is classified as an aggregated actor because his actions are not independent of his benefits and he is no longer member of an organized group. However, he is part of the general network of agriculture in Petorca.

Table 21: Alfredo. (Own research)

No.	Name (fictitious)	Sex	Age	District	Profession
30	Alfredo	M	65-80	La Ligua	Business Owner

MOTIVATIONAL AND COGNITIVE ASPECTS
Alfredo states not to be affiliated to any political party. He tells that his business depends on the agricultural sector. His main interest is his com-

---

46   For more information on water round table please read chapter 8.2

pany's growth and therefore the growth of the agricultural business. However, he claims to be a businessman »by accident«. He seems to be proud of being a self-made man and stresses his care about social equality. He repeatedly states that he originally was a low-paid social worker who worked in different projects all over Chile. He underlines that during several years of working with small-scale farmers as a social worker, he got to know this sector and the behavior of its actors very well. Furthermore, Alfredo criticizes the social distance between agricultural entrepreneurs and small-scale farmers. Trying to be a mediator, he took over the role of the director in the *water round table*. He reports to have sacrificed a lot of time and money on this work and finally resigned as he felt that people did not cooperate well. In specific, he expresses frustration about the individualistic society and about the lack of gratitude for his work.

The effects of the water scarcity described by Alfredo are the loss of agriculture, independent of its size. According to him, there are no socio-economic discrepancies with regard to water access as there are no more water rights to be bought and all users are struggling to satisfy their water demand. Alfredo does not confirm that the water scarcity is caused by climate change and reports that Petorca has always struggled with droughts. However, he believes that overexploitation of land and water for avocado plantations has led to the current water crises. Further, he mentions that the first government of Bachelet (2006-2010) promoted those plantations without recognizing the ecological consequences. In addition, he criticizes that the DGA lacks financial resources to control and sanction illegal water extraction. However, he does not believe that sanctions are a suitable solution. According to him, almost everyone in the Province owns illegal wells; therefore, a closure of those wells would lead to great unemployment. Alfredo refuses to give a detailed opinion about the water code and its connection to the scarcity. He declares to not understand the water law system. Nevertheless, he criticizes the separation of land and water rights because it allows water trade and speculations.

The most significant problem that impedes conflict resolution, according to Alfredo, is the lack of willingness to cooperate. This lack roots in an individualistic and opportunistic behavior inherent to the conflict parties and to the Chilean society. During the interview, he gives various examples for that behavior. He believes that Chilean society is historically divided, starting from the government of Allende. He tells that by then people had to decide if they were »revolutionist or nothing«. He says that those memories are very hard to overcome and impede reunions and cooperation until today. Alfredo points out that political ideology divides the conflict partners not only into small- or large-scale farmers but also into revolutionist or fascist, Moreover, according to him, personal interests often block resolution processes because some people take advantage of the conflict, for instance, in order to promote their own political career. He gives various examples of opportunistic behavior of the citizens, such as reporting water robbery while extracting illegally themselves and explains how this impedes solidarity and cooperation. In specific, he underlines that academics tend to romanticize small-scale farmers by underestimating their capacity of opportunism. He relates that under these conditions there is low willingness to take on responsibility and to cooperate. For example, he points to the never changing WUO leaders. At the same time, Alfredo believes that cooperation with the help of an interlocutor could help to overcome the conflict. For example, he underlines the importance of a *Junta de Vigilancia* and other forms of organization that unite small- and medium-scale farmers. Additionally, he states that small-scale farmers should prioritize the national market.

ORIENTATION OF INTERACTION
With regard to his orientation of interaction, Carasco gives an ambivalent picture. Firstly, he repeatedly claims to have a good relationship with everyone. This statement seems to base on his co-dependent interests and his identity as a businessman and counselor. As Alfredo says:

»[...] y eso le digo a la gente que viene a comprar, lo mejor que va a usted, lo mejor me va a mí.« (Interview 30)

(»[...] and that's what I tell the people who are buying here, the better it's going for you, the better it's going for me.« Translated by author.)

However, his close relationship to the LSAA and large-scale farmers should not be underestimated. He repeatedly demonstrates his anger about the opportunistic behavior of all actors involved in the water conflict. Although Alfredo was eager to promote cooperation, he finally gave up his position as a mediator and warns about the predominant individualism between actors. Taking that mistrust into account, but recognizing his remaining wish for cooperation, I classify his orientation of interaction towards all mentioned actor groups (*small-scale farmers, large-scale agriculture, civil society, governmental authorities*) as in-between selfish-rational ($U_x = X$) and cooperative ($U_x = X + Y$).

Part 4: Discussion

The first part of this study shows the current debate on water conflict and governance taking the example of Chile. Petorca Province is taken as a case study because it gives the opportunity to analyze in how far an institutional framework inspired by neoliberal principles impedes or facilitates sustainable water governance. Part 2 of this study describes how the research question was conceptualized, based on the socio-ecological system framework (SESF) by Ostrom (2007, 2009; McGinnis and Ostrom 2014) and actor-centered institutionalism (AI) by Mayntz and Scharpf (1995). The following chapters of Part 4 use the results shown in Part 3 as a base to test the hypotheses and to answer the research question stated in Part 2. They provide an evaluation of the solution capacity of the given institutional construct and propose strategies towards a context bound sustainable water governance.

8. The focal action situation: flows of water governance in Petorca

In the following, the theoretical model of AI and SESF was used to frame the empirical analysis of the water governance in Petorca. To test the hypotheses, process-tracing was applied (cf. Mahoney 2012; Vennesson 2008; chapter 4). By defining a context-bound sustainable water governance for Petorca, this study recognizes the complexity of the socio-ecological system and maintains a critical view. Questions of ethics, subjective socio-cultural preferences and its connection to the environment are considered.

The results of this investigation endorse that the governance style in Petorca is similar to the market style described by Pahl-Wostl (2015) (based on Meuleman 2008: 45-48) or minimal institutions by Scharpf (2006). The data indicate that the characteristics of the profound market-based ideology in Petorca prevent cooperation for higher welfare and distributive justice. In addition, it shows that several social innovations exist in Petorca and reveals the leverage points used by those projects to overcome the cooperation restricting mechanisms of the institutional framework. Finally, it discusses the potentials and limits of social innovation to pave the way towards sustainable water governance and concludes that although mechanisms to enhance cooperation are worth being supported, their effective radius is limited to certain actor groups. Consequently, deeper changes in the institutional framework are recommended. In specific, the paper suggests the governance architecture based on Ostrom (2007, 2009) and Pahl-Wostl (2015) called polycentric governance on river basin scale, which makes use of the before displayed leverage points and complements those mechanisms by combining governance modes and bottom up and top down strategies.

8.1 Market-based institutional framework

The first hypothesis states that the possible modes of interaction in Petorca are limited due to the institutional context. This is based on the assumption of AI stating that the institutional context allows or limits interaction to happen and influences the actors' solution-capacity (Scharpf 2006: 90f). Pahl-Wostl (2015) describes the classical forms of governance based on Meuleman (2008) as followed: market style, network style, hierarchical style. She states that, although no governance style could be explicitly categorized in one of these types, a classification is still helpful to describe the given institutional arrangements and to explain their inner logic (Pahl-Wostl 2015:91). According to Scharpf (1997), the categories of institutional contexts are anarchical fields/minimal institutions, networks, and regimes. According to Scharpf (2006: 62), it is crucial not to generalize but to precisely characterize in order to find accurate clarifications. Following Pahl-Wostl (2015) and Scharpf (2006), the data collected in the SESF indicate that water governance structure in Petorca is most similar to the market style based on Meuleman (2008:45-48) because it fulfils the criteria of control (by price), common motive (cost-driven), choice of actions (»free, ruled by price and negotiation«), power structures (»determined by degree of wealth and market share«), roles of government (»delivers services to society«) and roles of knowledge among others. Those features are discussed with regard to Petorca in the following paragraphs.

A WEAK STATE ALLOWS THE MARKET TO RULE
My findings show, the government in Chile has weak legal and financial power regarding to water governance and takes on a role classified as service delivery according to Meuleman (2008) rather than that of a 'Leviathan' (coined by Hobbes, cf. Ostrom 1990: 11). First, the price is seen as the regulator for the most efficient water use since water privatization

started during dictatorship in 1981 and has subsequently intensified. This means, the resource is treated like a good, and its shares can be bought and sold on the market (cf. chapter 7.4).

In the Chilean discourse the term privatization is criticized by some scholars such as the lawyer and specialist for water rights Alejandro Vergara (Vergara Blanco 2014: 89). However, in this study the term privatization is used referring to Chile because of the following reasons: While it is true that the water code declares water as a national good[47] and the state remains with the rights of expropriation in case of extreme catastrophe (e.g. drought) (CA art. 27), I argue that by handing over water property to private entities, the water code, the Chilean constitution and the implementation of the law treats water as a private good in a free market.[48] The same article of the water code that declares water as a national good continues as follows:

> »[...] y se otorga a los particulares el derecho de aprovechamiento de ellas, en conformidad a las disposiciones del presente Código.« (Código de Agua, 1981, Art. 5)

> (»[...] and the rights of exploitation of these [waters] will be granted to private entities, in approval of the dispositions of the present code.«)

Furthermore, the Chilean constitution in article 19, No 24 says:

> »Los derechos de los particulares sobre las aguas, reconocidos o constituidos en conformidad a la ley, otorgarán a sus titulares la propiedad sobre ellos«

> (»The rights of the private individuals over the waters, recognized or constituted by law, grant to the title holder the property over these«)

By that, the water code and the Chilean constitution turn water into a private good and make the first part of Article 5 (CA 1981) that declares water as »national goods of public use« obsolete. In addition, no legal priority for drinking water exists (Rivera et al. 2016: 40), and the right of expropriation has never been used – not even under severe environmental conditions, such as the so-called mega drought during 2010-2015 (cf. Garreaud et al. 2017 and chapter 7.3).

Further, the legislative and executive power of controlling and sanctioning water access and distribution and water rights in Chile is strictly limited. In 2005 a more democratic and sustainable water code reform was in-

---

47 »Waters are national goods of public use [...]« translated by author (Las aguas son bienes nacionales de uso público [...]) (Código de Agua, 1981, Art. 5)«
48 For deeper insights into the Chilean law system please see chapter 7.4 [GS6 rules in use].

troduced demanding inter alia enhanced rules for obtaining water rights. This water code reform, together with other institutional steps taken since, was criticized mainly for being too weak (OECD/ECLAC 2016; Retamal et al 2013; Valdés-Pineda et al 2014, Larraín 2010, cf. chapter 7.4). Moreover, the high diversity of governmental bodies directly or indirectly influencing water governance weakens the state's power and shows a tendency to overlapping responsibilities, competition, and complex bureaucratic regulations. On the provincial level of Petorca, 23 institutions of public or semi-public structure were found in this study (cf. chapter 7.4). Several civil servants interviewed complain about high bureaucracy, lack of transparency and competition that is produced by the high number of water institutions as well as their distant relationship to society. They report that this »chaotic« structure weakens the state's power to monitor and allows illegal water withdrawal to occur (Interviews 13-16). Moreover, water users are discontent about this problem and complain about institutions working parallel instead of complementing each other, which leads to confusion and impedes development (cf. Interview 12). The criticism focuses mainly on the National Water Directive (DGA). In public and in private interviews, this institution has declared itself as incapable of complying their task because of insufficient human resources and legal power. For example, concerning human resources, an employee of at the regional DGA office criticizes that only two staff members are responsible for monitoring the compliance of water right rules for the entire Vth region[49]. His main criticism is that:

- firstly, the DGA is not allowed to start investigating by its own but depends on an initial report of a citizen,
- secondly, if irregularity has been reported, the DGA must call the accused person and ask for permission to enter the property (Interview 35).

The weak power of the DGA was subject of criticism by several interview partners of all actors' groups (while only by one person of the actor group *large-scale agriculture*). One of the small-scale farmers demonstrates his discontent and introduces another caveat as he states:

---

49 The Vth region called »Region de Valparaíso« is the administrative division to which Petorca belongs. The Vth region has an area of about 16400 km2 with a population of approximately 1,8 million people.

»O sea las leyes son muy débiles, por decirte, si un usurpador de agua está robando el agua y lo pilla la DGA le dice `a usted le voy a pasar una multa' –'ya de cuanto es'-'200 mil pesos'- 'a ya la pago' y sigue robando los ríos. Entonces ¿tú crees que eso es justo?« (Interview 7)

(»The law is very weak, for example, if a water user is stealing water and gets caught by the DGA. They say, 'I will charge you a fine' – 'well, how much?' – '200 thousand pesos' – 'ok. I will pay immediately.' And he keeps on stealing from the rivers. So, you think that this is fair?« translated by author.)

Currently, the highest sanctioning possible according to the water code is a fine of max. 20 UTM (approx. 1260 €).[50] Though I concede that this fine is higher than in the example given in the statement, I still argue that it can be compensated for by the firms wins and therefore, adds to the argument of weak power.

Moreover, this situation of weak legal power seems to put pressure on the DGAs civil servants. For example, a civil servant reports to be sued personally for reporting water robbery (Interview 22) and activists as well as municipal workers report that DGA members and other responsible people were removed from their positions after doing a profound research on illegal water extraction in Petorca (field observation; Interviews 5, 14). These statements could not be proven during this study. A leading member of the DGA says in an interview that every member left their position out of free will (Interview 22). Nevertheless, the statements underline the weakness of the institute because employees who are threatened to be personally sued tend to be less capable of dealing with critical situations as personal costs are at stake. In addition, the DGA is criticized by different interview partners, media and scholars for mismanagement because the water rights for the Rivers Petorca and Ligua have been over-assigned although this area was declared »zone of catastrophe« (Interviews 14, 18, 23, 28). These conditions lower the motivation of society to report water robbery and, therefore, weaken the state's control over water management. For example, different interview partners stated to be hopeless that any legal dispute would lead to a satisfactory outcome because of the power differences that come along with different financial budgets (cf. chapter 7.5.1; Interviews 3, 4, 6, 12).

The criticism displayed is confirmed by the OECD. As being a member of this international organization, Chile is encouraged to meet their standards in water management. The OECD reports state that there have been

---

50 Here calculated on the 26th of November 2018.

improvements since 2005 (OECD 2005, OECD/ECLAC 2016). Nevertheless, current OECD recommendations demand the development of measurement and monitoring strategies of water quality by setting quality standards for all rivers and groundwater bodies and the improvement of water management by prioritizing human consumption and sanitation, strengthening market transparency and setting effective and enforceable abstraction limits to stay within ecological and social limits (OECD/ECLAC 2016: 78). The above text describes the state's limitations of monitoring and sanctioning. Consequently, its role shifted over to a service consisting essentially of counseling, providing subsidies and assigning water rights.

UNEQUAL POWER STRUCTURES: EDUCATION & FINANCIAL RESOURCES DECIDE ABOUT ACCESS TO THE GAME

As already mentioned, in this context of minimal institutions the price remains the controlling factor. Consequently, I argue that power structures and decision-making processes in water governance are determined by the degree of wealth, market share and information of the individuals (cf. Pahl-Wostl 2015: 92). As described by Pahl-Wostl, in an idealized market

»[...] individual actors do not have any powerful position. There is no direct interaction between actors – they only interact via the institution of the market accessible and visible to all. [...] The 'power' of steering resides in the institution of the market itself. Real markets are not ideal and some actors are more powerful than others (Pahl-Wostl 2015: 89)

In Chile, ideally, everyone should be able to participate in the water market by obtaining water rights. However, financial power and information are required to participate and to obtain water rights. Considering data on socio-economic factors in Petorca, not all actors are of equal power. In addition to the low financial power and low educational background of most Petorca residents (cf. chapter 7.1), a significant lack of information impedes farmers to participate in the market. This observation was noticeable during meetings with small scale farmers and underlined during different interviews (cf. observations April 6, 2016; Interviews 5, 14, 20).

Contrary to an idealized market, formal local organization of water governance, here called Water Users Organizations[51] (WUOs), do exist (cf. chapter 7.4). However, the WUOs' power structure is determined by the market share of its members since each member's vote counts according the water rights they possess (Water Code Art. 222). A WUO in charge of monitoring the water right rule compliance, called Junta de Vigilancia (JV), has not yet been installed.[52] According to field observation and expert interviews, there are basically two problems of installing a JV: (a) small-scale farmers oppose to build a JV because because they fear to be outvoted by larger water right holders; (b) the current situation of regularized property rights in Petorca is chaotic. An expert estimates an unofficial number and says that about 4000 water rights are not correctly registered (information received from PGH member). In addition to family conflicts and conflicts in small community, this situation of ambiguity leads to bureaucratical obstacles whenever water right owners want to join governmental programs (e.g. JV, CASF, CAST etc.), enter the water market or need to justify water use. The water right regularization process is costly, which again turns financial wealth into one of the determining factors of power in water governance. This water right clarification is supported by a program called »Programa de Gestión Hídrica« (PGH) financed by the Chilean government. It aims at facilitating market participation and provides access to governmental subsidies for water right clarifications. But, even if property rights are clarified, the monitoring of related extraction rights remains cost-intensive because of the spatial distance to the monitoring authority DGA or consulting fees, for instance. In practical terms, a person of higher financial power has more options to illegally withdraw water since he or she can afford more expensive techniques, on the one hand, and may drain a considerable amount of water undisturbed, on the other hand.

The cognitive aspects of the interview partners contribute to the argument that the financial and educational background determines the power structure of water distribution and access. A pertinent illustration is provided by various actor groups. Water activists criticize this market-based

---

51 A WUO can be built by two or more persons who possess water rights of the same water body (river, groundwater, canal etc.). Their main management tasks are to distribute water, administrate constructions and resolve conflicts as well as trade of water rights. (cf. chapter 7.4, MOP and DGA 2015: 120; Retamal et al. 2013: 13)

52 More information about different kinds of WUOs is to be found in chapter 7.4.

style of governing water. Besides too low governmental power, they point to the win-maximizing motivation of this system (Interviews 5, 7, 8). As one activist concludes:

> »Pero además esta es una institucionalidad que se encuentra captada por el modelo de lucro con el agua, es una institucionalidad que se señala que en Chile el problema de las aguas no es un problema de propiedad, sino un problema de gestión y de libre mercado, es decir son autoridades que se encuentran absolutamente convencidas de que el agua es un recurso económico, inagotable y que ese recurso económico inagotable debe darles beneficios a quienes son sus propietarios.« (Interview 8).

> (»But in addition, the institutions are led by the profit model when it comes to water, the institutions point out that in Chile the problem of water is not a problem of property but a problem of management and free market; in other words the authorities are fully convinced that water is an economic good, endless, and that this endless economic good should benefit those who own it.« Translated by author.)

Contrary to his belief, the representatives of the governmental water institutions and municipalities interviewed in this research indicate the downsides of the market-based management system. To give a concrete example, Ramón thinks that inequality in Chile is produced by the neo-liberal structures, not only in water management but in different political areas. He complains that this reduced governmental power complicates his work:

> »Y eso es un tema cultural nuestro como país. Un tema de estructura económica también, ¿cierto? Un apego a las normas del mercado muy profundo, me imagino yo, dónde el estado tiene una participación pequeña. Cómo lo hablamos de la DGA. No tienen los recursos suficientes. Yo aquí trabajo con 10 personas. Son mi equipo y tiene que ver todo. […] o sea, un equipo reducido. Un estado reducido. Sin alegar. Sino que, estos son las condiciones. Entonces, si uno quiere llegar a aquellos que son más vulnerables, cuesta más, po.« (Interview 17)

> (»And this is an issue of our nations' culture. An issue of economic structure, also. A very profound closeness to the market rules, I believe, in which the states participation is very low. As we talked about the DGA. They don´t have enough resources. I work with 10 persons here. We are one team and we must look after everything. […] well, a very small team. So, if you want to help the most vulnerable, it is even harder.« Translated by author.)

Other actors from the group of governmental authorities find fault in the distribution and access of water as the system favors people with higher financial resources (Interviews 17, 18). Further, Gabriel points out how this inequality influences the water management structures:

»Él que tiene la mayor cantidad de derechos - eso se convierte en acciones, él va a definir la política en las comunidades de agua. Y ya decimos que los distintos derechos de agua han sido entregados, y a simple vista vemos que los pequeños agricultores tienen muy pocos litros por segundo, tienen muy pocas acciones. Entonces, [...] el poder de decisión dentro de estas comunidades de agua es escaso.« (Interview 5)

(»The person with more rights – which will be transformed into shares – will define the politics in the water communities. And we already said that several water rights have been delivered, and at a glance we see that small farmers receive very few liters per second, they own very few shares. So, [...] the power of decisions in these communities is deficient.«)

Moreover, this perception of unequal power structures is demonstrated by several statements, especially of the actor group civil society (cf. chapter 7.5.1). In this study, unequal power structures are discussed explicitly in chapter 8.3

WHAT ABOUT OTHER INSTITUTIONAL CONTEXTS?
The above described features of common motive, choice of actions, power structures, roles of government and roles of knowledge are all in line with market-style governance. However, other governmental styles and hybrid forms exist in theory and are compared to the research case in the following paragraphs.

Even though, a high number of state authorities exists that directly or indirectly influence water management, the dominant power stays with the market and its shareholders (cf. chapter 7.4). Consequently, there is no (centralized) authority with sufficient power to implement a mainly hierarchical governance style in the case of Petorca. Another type of classical governance style is the anarchical field. As in Petorca rules and regulations e.g. for property and withdrawal of water exist, the case cannot be categorized as being characterized by an anarchical management style. However, it should be noted that even though those regulations exist, their implementation seems to be highly deficient. Furthermore, the actors are in the relationship of a local network, as they »*jointly contribute to the economic vitality of [...] [Petorca] (Hull/Hjern 1987; Sabel 1989)*« (Scharpf 1997: 156) However, the actors are not in a network concept in the sense of the classic governance styles described by Scharpf (1997) and Pahl-Wostl (2015), insofar as these would imply trust as the controlling factor, recognition of the government as a partner and a high degree of self-organized coordination. This »cooperative quality« as a defining feature seems to be missing in Petorca (ibid.) and will be analyzed in detail below.

AI distinguishes between anarchical field, network, association, and hierarchical direction (regime) (cf. chapter 5.2, Scharpf 1997). The market style is not part of this list. Nevertheless, it will be used here as it suits to the water governance style of Petorca most.

8.2 Patterns of interaction orientation

The first hypothesis states that interaction modes in Petorca are limited due to the institutional framework. Hence, to verify that correlation, modes of interaction must be filtered in order to determine their variety. The following subchapter discusses modes of interaction in Petorca. First, it shows that cooperation seems to be generally low. Then examples of different big- and small-scale projects are discussed with regard to their cooperative quality. On the whole, while big scale projects have a strong tendency to fail, small scale projects show a higher degree of success and cooperation. These small-scale projects are called social innovations (as defined in chapter 3).

In line with the institutional framework of minimal institution respectively market-style governance, trust is low and possible action modes according to the AI are limited to one-sided or mutual adjustment and negotiated agreements (cf. chapter 5.2.4). The latter is most likely restricted to »spot-contracts«. This means that there is neither institutional nor intrinsic compliance assurance so that the involved actors only agree or disagree on minimal steps. The number of such exchange processes is limited and welfare maximizing gains may be lagging behind. According to Scharpf, these losses could be avoided if the negotiation allowed profiting actors to compensate their disadvantaged counterpart (Scharpf 2006: 214). He uses »*distributive bargaining*« as a technical term to describe this mode of interaction (ibid.). To implement distributive negotiations, common welfare needs to have a higher priority than personal interest. However, in the institutional frame of market mode, this premise seems not to be given because of competitive or rational self-interested orientation of the actors.

In the following paragraphs, I establish if and how far the interaction modes in Petorca are restricted to one-sided adjustment and limited negotiations. In order to do so, the cooperative quality of different organizational structures and collaborative projects is estimated by way of using the qualitative and quantitative data collected by implementing SESF and the interaction orientations discovered by using the AI (cf. Part 3).

8.2.1 Individualistic behavior, competition and low cooperation

To estimate the organizational quality of the formal, but also of informal interaction in Petorca interviews and field observations are taken as a base. Considering the orientations of interaction found in chapter 7.5, it can be said that selfish-rational interaction orientation dominates. This means, that actors are mostly interested in their own profit $(U_x = X)$[53]. Thus, they do not consider the gain or loss of another group as contributing to their own gain or loss. The actor group receiving most selfish-rational orientation is *governmental bodies and authorities*, followed by *activists*. Interestingly, cooperative orientation of interaction – meaning that the gain and loss of others is added to the personal profit $(U_x = X+Y)$ – was detected as the second strongest orientation throughout the interviews. Most cooperative orientation is shown towards the actor group of *civil society*. The dominating presence of both cooperative and selfish-rational orientation may seem contradictory. However, this phenomenon can be explained considering the subjectivity of the interview partners on the one hand, and the context of the interaction orientation on the other hand. Both arguments can be demonstrated by an example. While almost every one of the actor group *civil society* shows cooperative orientation towards other WUOs and community members, they criticize that most members of *civil society* behave individualistic or selfish-rational. Hence, they see themselves as outstanding examples of their own actor group. This displays the former argument of subjectivity. The latter argument of context-bound behavior can be explained by the actor group of *regional authorities*. Here, cooperative and selfish-rational interaction orientation is almost equally shown. However, it depends on which actor group this orientation is directed to, *regional authorities*, for instance, show cooperative orientation towards *large-scale agriculture* and a selfish-rational orientation towards other *governmental bodies and authorities*. The following Table 22 provides an overview of interaction orientation and their direc-

---

53   U_x refers to the actors (ego) estimated utility. X is egos gain and Y is alters' gain (cf. chapter 7.5)

tion. However, it should be underlined that the data shown are qualitative and therefore, only and at most, indicating tendencies.

Table 22:Orientation of Interaction between actor groups in Petorca Province. (Own research)

Towards / Group	Large-scale agriculture	Governmental bodies/ authorities	Activists	Civil society	Water provider
Local leaders					
Activists					
Citizens					
Local authorities					
Regional authorities					
National authorities					
Large-scale agriculture					
Water provider					

Hostile — Competitive — Selfish-rational — Cooperative — Not mentioned

Orientation of Interaction

INTERACTION GAPS

Further, the interviews reveal that most interactions take place between the actor groups of *governmental authorities* and *civil society,* and between *governmental authorities* and *economy* (large-scale agriculture and water provider). However, the former interaction seems to be one-directional, and the latter is mostly concentrated on *regional* or *national authorities* as the relationship on a local level, especially towards representatives of large-scale agricultural companies, seems distant. Direct interaction between representatives of those companies and *civil society* is rather limited. The qualitative data indicate an interaction gap between these two actor groups. Formally organized cooperation, in *Water Round Tables* for example, has failed. In WUOs *large-scale agriculture* is mostly seen as an obstacle that needs to be handled instead of a potential partner for cooperation (cf. chapter 7.5.1).

Additionally, informal interactions are generally limited to complaints by *civil society* about *large-scale agriculture*. Activists claim to have no or very little contact to the actor group of *large-scale agriculture*. As a consequence, the dominant arenas of interaction between *large-scale agriculture* and *civil society* seem to be employment and legal disputes. With regard to the interaction internal to the actor group, the cooperative quality varies.

The actor group of *governmental authorities* is repeatedly criticized for an internal lack of cooperation by their own members and by the actor group of *civil society*. Interviewees report competitive behavior and lack of transparency that lead to a lack of communication and unclear responsibilities (Interviews 12-15, 22, 29). While on a local level the cooperation seems to be higher, the criticism especially targets at cooperation on regional, national and crossing levels. The actor group of *large-scale agriculture* seems to have a higher cooperative quality. It is connected via the large-scale agriculture association and mostly cooperative among each other (cf. chapter 7.5.3).

The most significant lack of cooperation seems to be within the actor group of *civil society*. Most interviewees define lack of interest in organization as a mayor problem for civil society and water management in Petorca (only eight out of 33 do not mention this topic). Almost all interview partners who live in Petorca, such as *small-scale farmers, activists, local leaders* and *local authorities,* criticize the lack of organization in civil society. Several of them complain about the problem to motivate people to cooperate (cf. chapter 7.5.1). Furthermore, most representatives of *regional and national authorities* point to this topic (cf. chapter 7.4.2). Farmers of the actor group *large-scale agriculture* are the only ones that do not mention the lack of organization.

Since the actor group civil society is the largest in terms of numbers, and its lack of cooperation seems to be outstanding, the cooperative quality of that actor group and the peculiarities of the formal cooperation need to be discussed in more detail. The most numerous forms of formal and informal cooperation in water management in Petorca are WUOs. The WUOs can be divided into two groups. The first group is WUOs which members are water right holders and whose water rights are for industrial use. This group splits up in rather small WUOs consisting of two AC, 88 CASF and 12 CAST (cf. chapter 7.4). Technically, water right holders of one water body (such as a canal or ground water aquifer) in Petorca can participate in decision making processes of water governance via those organizations. Such decisions may concern issues like the construction or maintenance of water infrastructure, participation in local events or voting for WUO leaders. The second group is associations of rural potable water (APR). In the Petorca, La Ligua and Cabildo Districts, 36 official APR exist. Further, there is an indefinite number of unregistered APR. For example, a report of the University of Playa Ancha counts 15 informal APR out of 25 in the Petorca District in 2014 (UPLA 2014). The basic task of an APR is to assure domestic water supply in rural areas. APR members work voluntarily on administration, operation and maintenance of water infrastructure. Members of the individual WUO and experts working with those WUOs point to low participation (Intervies 14, 29; cf. chapter 7.5.1). They criticize that always the same persons and mainly elderly citizens participate. Significantly, occupation of leadership positions would not change. A major problem mentioned by WUO members, governmental representatives and other experts is the insufficient willingness of WUO members and other citizens to take on responsibilities, for example, for common activities like cleaning channels, or for other members by taking on leading positions (Interviews 1, 11, 15, 19, 29, 30). Instead, WUO members prefer individual solutions. For example, there are several complaints about a lack of maintenance of the channels because people rather care for their private wellbeing instead of sharing the responsibility for the channel cleaning. In addition, experts certify that competitive behavior amongst different WUOs impedes cooperation (Interview 14).

8.2.2 Cooperative quality of collaborative projects in the field

The qualitative data above discussed concentrate on the actors' perception. To overcome the factor of subjectivity, in the following paragraphs the cooperative quality is analyzed based on real water governance decisions in the field. This procedure aims at testing the perception of the actors with the observed structures and at estimating the cooperative quality in Petorca water governance. Therefore, the interplay of the interaction orientations and the modes of interaction is examined on collaborative projects aiming at alleviating water scarcity and conflicts.[54] The projects discussed can be divided into (a) large-scale projects over-reaching various communities and actor groups, and (b) small-scale projects mainly concentrating on one or a few actor groups and communities. Both were selected from the results shown in Part 3. The discussion shows that especially large-scale projects which need complex negotiations, majority votes, or hierarchical decisions, are not successful in Petorca independent of the nature of the strategy applied, be it technical, organizational or juridical. Social innovations, in contrast, open space and foster new modes of interaction. Hence, they show more cooperation quality. However, one large-scale project seems to be an expectation to this phenomenon.

The projects are listed in Table 23 and described with the focus on their modes of action[55] and implementation barriers in the following paragraphs.

TECHNICAL SOLUTIONS

Chapter 7.1. of this study displays the current and upcoming technology for improving water management. The building of water tanks and a desalinization plant has repeatedly been discussed in the Petorca Province. The planned water tanks in *Las Palmas*, *Los Àngeles*, *El Pedernal* and *La Chupalle* proposed by the National Water Directive (DGA) and Directive for Hydraulic Works (DOH) have not been built yet because small-scale farmers oppose to the construction processes. They believe that cases of corruption can occur in connection with the construction of water tanks and that they will suffer damage because the water flow will be altered. Furthermore, they suspect that the accumulated water will primarily serve big water right holders (large-scale agricultural companies). Governmental bodies have commissioned implementation studies for several water tanks. The next step following the estimation of the costs is to find a construction company. However, only one project has completed the public tendering procedure so far (water tank Las Palmas). The decision-making process for the project implementation is basically hierarchical.

---

54 Private projects such as improvement of irrigation systems are not in included because they do not provide information about the interplay of actors neither of various actor groups.

55 Modes of action are ways of social action coordination reaching from uniliteral action, over negotiated agreement and majority vote to hierarchical direction (cf. chapter 5.2.4)

Table 23: Large-scale projects aiming to solve water scarcity in Petorca. (Own research)

Strategy	Technical	Organizational	Juridical
Description	Implementation of large-scale water infrastructure	Creation of new participatory or self-organizational practices	Creation or Modification of the Law
Project	**Water Tanks**	**Junta de Vigilancia**	**»Ley del Mono«**
Scale	Cross-community	River basin (cross-community)	Provincial
Actor-groups	Small-, medium and large- scale farmers, governmental bodies, APRs	Small-, medium and large- scale farmers, governmental bodies	Small-, medium and large- scale farmers, governmental bodies
Description	Four water tanks are planned by the government to accumulate mostly irrigation water	WUO installed by the DGA and self-organized by water-right holders to monitor water use	Modification of Water Code to allow legalization of informal wells
Project	**Desalinization Plant**	**Programa de Gestión Hídrica (PGH)**	**Water Code Reform**
Scale	Provincial	Provincial	National
Actor-groups	Municipalities, civil society, water provider	Small-scale farmers, governmental bodies, academia	All actor groups involved
Description	The mayor of La Ligua started a pilot project for a desalinization plant to provide the province with drinking water	Program financed by the Ministry of Agriculture and organized by the PUCV providing technical and legal consulting services	Water code reform aiming social and environmental sustainability.
Project		**Water Round Table(s)**	
Scale		Provincial	
Actor-groups		Small- and large-scale Farmers, Public Institutions	
Description		Round tables to facilitate conflict resolution	

Nevertheless, authorities seek the society`s approval to avoid conflict. For example, they inform WUO members and open space for questions during public meetings.

Another technical solution discussed in Petorca is a desalination plant. The project is mainly supported by the Ministry for Public Works (MOP; Gobierno de Chile 2010) and the municipality of the province capital La Ligua. A prototype of a desalinization plant was inaugurated in 2014. However, official data for this project are scarce. A document from MOP states that there are several plants planned for the Province that will provide drinking and irrigation water (Zarricueta Carmona 2015). A representative from the municipality of La Ligua underline that such a desalination plant could alleviate water stress not only for the La Ligua District, but also for Cabildo and Petorca (Interview 16). While the municipality of

Cabildo supports that idea (Interview 13), most interview partners do not. They believe that it is too expensive (Interview 1, 26) and ineffective (Interview 4). Moreover, activists suspect that it is just a strategy to silence the conflict and to cover up illegal water robbery (Interview 5). Additionality, it is important to note that there is no legal framework yet to install such a plant, as the Water Code applies for inland surface and ground water but not for sea water.

ORGANIZATIONAL SOLUTIONS

The *Junta de Vigilancia (JV)* is repeatedly named as an organizational solution strategy by members of different actor groups (activists, governmental bodies, etc.). This type of WUO would allow water right holders to cooperate on a greater scale as it aims at monitoring and sanctioning norm compliance of users belonging to one river basin (e.g. river basin La Ligua or river basin Petorca). Approximately since 2010, efforts have been made to prepare the bureaucratic steps of the formation of a JV in the basin of Petorca River. One of the main problems is the unresolved legal situation of a great number of water users. A program financed by the state that supports water right holders to clarify their legal situation has been running since 2010 in Petorca. Efforts have mostly concentrated on the Petorca River basin so that different water right holders of that basin have already prepared to take on legal positions in a JV. Nevertheless, the obstacles seem to be so high that no JV currently operates in the Petorca Province, neither in Petorca nor La Ligua River basin. Besides the chaotic legal situation of water rights, another problem exists: As the JV is composed of the water shareholders themselves, a majority agreement is needed for project execution. Small-scale water right holders reject the concept of JV because they fear to be disadvantaged in its decision processes.[56] Thus, they block the process of building a JV.

The second example for an organizational solution is the *Programa de Gestión Hídrica (PGH)*. This program aims at finding new solutions to cope with the drought in the agricultural sector. It was founded by the Pontificia Universidad Católica de Valparaíso in 2015. PGH's office is based in the city of La Ligua, and its multidisciplinary team consists of

---

56 The voting system in a Junta de Vigilancia is bound to the amount of water rights hold by their members. This means, that a member's vote weights in accordance to their ownership. Hence, each vote does not have the same weight, but the weight correspond to the amount of water rights owned (cf. chapter 7.4).

agronomists, lawyers and economists. Its target groups are small- and medium- scale farmers all over the Petorca Province. Basically, PGH seeks to fulfill three goals. The first aim is to provide education about water rights for small-scale farmers. The second goal is to implement a free of charge program to regularize unclear legal situation of water rights for small-scale farmers. The third aim is to test and develop new crops and products starting from cultivation to marketing. In contrast to other large-scale projects, the PGH has already been implemented in the province and has succeeded in increasing cooperation with different actor groups. Although the organizational structure of PGH is rather hierarchical (e.g. strategic decisions must be agreed upon with the PGH head and an annual report must be written to the PUCV), decisions concerning the community directly are made in small groups including one or more medium- and small-scale farmers, and members of the *PGH*.

Furthermore, diverse *Water Round Tables* were installed in Petorca as a response to the drought and its ongoing conflicting consequences. However, data about the development and organizational structure of those *Water Round Tables* are limited. According to the interviews and field observations, the first *Water Round Table* on a provincial scale was introduced approximately in 2008 (Interview 30). The *Water Round Table* intended to facilitate decision making in water governance and to contribute to conflict mitigation and solution. Therefore, it aimed at the participation of both, public and private entities. The leadership of this *Water Round Table* was assigned to Alfredo, leader of the association of medium- and large-scale agriculture in Petorca (LSAA). After the *Water Round Table* stopped working in 2013, Alfredo decided to leave the asscociation. He left both positions because he felt that cooperation between the different actors is not possible and participants would try achieving their personal goals instead of seeking common solutions to the water scarcity (Interview 30). In 2014, the government created the office of *Delegado Presidencial de los Recursos hídricos* that was commissioned to install a *Water Round Table* (*mesa de agua, or mesa tecnica*) on a provincial scale that aimed at the same goals as the *Water Round Table* led by Alfredo. The *Water Round Tables* are criticized for similar reasons as the WUOs. Activists and small-scale farmers criticize that the *Water Round Table*, specifically from 2008 and onwards, was led by the large-scale farming industry who manipulate these meetings for their personal goals (cf. Interview 5).

JURIDICAL SOLUTIONS

With regard to juridical solution strategies, two projects are important to this study. First, in 2005 a modification of the Water Code was made, commonly called *Ley de Mono*, which has influenced the discussion about possible conflict resolutions in Petorca until today. This modification aimed at legalizing informal underground water extraction. Well owners got the possibility to regularize water rights for up to two liters per second for wells constructed before June 2004. Although the modification targeted at small-scale farmers by handing out water rights regularization funds, the process was open to all water right holders. Consequently, it turned out that small-scale farmers struggled in the assignment process due to the unclear legal situation of their water rights, and many large-scale farmers took advantage of being assigned large number of wells. Therefore, the DGA decided that the *Ley de Mono* could lead to an overshare of water rights and stopped the process (Budds 2012). Some interview partners suggest a second try of an improved version of this modification. For example, the representative of the LSAA Javier suggests *Ley de Mono* as a solution strategy to overcome the great amount of legally unclear water rights in Petorca (Interview 25). However, other interviewees point out that this law was abused by large-scale farmers (Interviews 23, 30). Some may argue that the *Ley de Mono* was successful as it reached implementation. However, after the implementation it did not reach the intended goals as it rather reinforced water access inequality instead of balancing it. The DGA recognized this problem and stopped the project shortly after the implementation.

The second juridical solution strategy important to the case is the currently discussed *Water Code Reform*. The *Water Code Reform* includes inter alia prioritization of water for human consumption and an end of infinitive property rights. Several small-scale farmers and activists are in favor of that reform and have discussed it with their community and politicians. However, although the reform passed the congress vote in 2016, it needs to be approved by several committees and finally by the senate to be implemented. To date, these steps have not been completed.

The examples of large-scale projects here displayed turned out to be the most discussed projects during the field research. Except for PGH, none of them was (successfully) implemented. The data indicate that modes of action are limited and concentrated mainly on one-sided adjustments as in the situations described here, complex negotiations, majority votes or hierarchical decisions were not (successfully) made.

8.2.3 Social innovations in Petorca

Moreover, several implemented and on-going medium- and small-scale collaborative projects were found that directly or indirectly aim at a solution of the water scarcity and conflict. Those social innovations broaden the action modes insofar as their cooperative structure requires a more complex interaction than one-sided adjustment or limited negotiations. The following Table 24 displays a selection of small- and medium scale collaborative projects. They are categorized with regard to their solution strategy for water scarcity and conflicts.

Table 24: Social innovations (SI) in Petorca Province. (Based on TEPSIE (2014: 15) and Scheidewind et al. (1997) and own research)

Cate-gory	New services and products*	New participatory /organizational practices*	New knowledge production*
Descri ption	Creation of new services or products	Creation of new participatory or self-organizational practices	Creation of new knowledge production processes
SI	**petorQuinoa**	**Minga por el agua**	**Oficina de Asuntos Hídricos**
Scale	Provincial	Community	Community
Actor groups	Small-scale farmers, governmental bodies	Citizens, governmental bodies	Small-scale farmers/ citizens, governmental bodies, academia
Descri ption	Farmer cooperative promoting quinoa (andean plant with low water demand)	Self-organization of eight households (mainly widows) to install water pipeline	Joint office for communitarian work related to water by the municipality of Petorca, UPLA and Rural Water Association Union (Union de APR)
SI	**Banco de Semillas (Seed Bank)**	**MODATIMA**	* categories build on the bases of TEPSIE (2014:15) and Schneidewind et al. (1997) **MODATIMA started as a provincial project but widespread nationally
Scale	Community	Provincial**	
Actor groups	Small-scale farmers, NGO, governmental bodies	Small-scale farmers, citizens	
Descri ption	School project that aims for a joint protection and share of crop seeds	Activist's organization to fight against privatization of water rights	

NEW SERVICES AND PRODUCTS

The projects *petorcaQuinoa* and *Banco de Semillas* both follow a solution strategy that offers new types of services and products. *petorQuinoa* is a farmers' cooperative for quinoa plantation. Quinoa is an Andean plant that requires comparatively little water. The initiative started in September 2015 as a solution strategy to the ongoing water scarcity. In search of an alternative to the water requiring avocado production, farmers decided to promote quinoa. Originally, *petorQuinoa* started in cooperation with 44 small- and medium scale farmers from all over the Petorca Province, together with the Chilean institute for business development CORFO and an agriculture consultancy agency (ConsultoriaAgricula). In 2017, the initiative lost one member. However, 20 small- and medium scale farmers are on the waiting list to join the cooperative (Interview 31).

The project *Banco de Semillas* evolved in March 2016 and is organized by the Liceo Técnico Cordillera of Petorca (Technical College Cordillera) in cooperation with the University of Playa Ancha, the municipality of Petorca (Oficina de Asuntos Hídricos) and the Italian NGO Progressio. The innovation emerged in reaction to a loss of traditional crops and as an answer to the drought. The main idea is to multiply ancient seeds and support small-scale farmers of the region.

NEW PARTICIPATORY AND ORGANIZATIONAL PRACTICES

Organizational initiatives that differ from the formal WUOs formerly mentioned were found on a smaller scale. *Minga por el Agua* is an initiative of different local actors that helped to install a water pipeline to eight houses in the district of Petorca. The installation was coordinated with the help of the municipal office for water issues (Oficina de Asuntos Hídricos) and implemented with the collaboration of different APRs. Traditionally, the word *minga* is derived from the communal work of indigenous people in the Andean region. In Chile, these forms of community work, in which one family would host a work party accompanied with food and drinks, are known from the Mapuche tradition (Andolina et al 2009). *Minga*s are still popular until today and mostly found in the southern region close to Chiloé. Although the installed water connection is not suitable for drinking water, it alleviates the water scarcity as the water can be used for other household needs (irrigation, cloth washing etc.).

Another organizational phenomenon concerning water governance to be mentioned is the activist group MODATIMA. Originally, the group was founded by small-scale farmers and workers of the agricultural business from the three districts Petorca, Cabildo and La Ligua. They have organized campaigns against water privatization and reported illegal water extractions by large-scale agricultural companies in Petorca since 2010. Although some formal structures like president and secretary exist MODATIMA considers itself to be an informal movement that is developing constantly, rather than an NGO or association (Interview 34). Various MODATIMA groups were founded all over the country, and they recently carried out campaigns and collaborations with the national and international media. Because of its growth, MODATIMA stands out from the group of small- or medium scale projects because of its regional scale and the high number of actor groups involved. However, it differs from the formal organizational structures formerly mentioned and started as a small-scale project mostly developed form members of the actor group civil society. Therefore, it is treated as social innovation in this study.

NEW TYPES OF KNOWLEDGE PRODUCTION

The third category consists of new types of knowledge production. The project is called »Oficina de Asuntos Hídricos« (OAH, Office for water issues) and started in 2016 as an initiative of the municipality of Petorca, the Universidad de Playa Ancha and the Fundación de Asistencia Internacional (FAI) (UPLA 2015) The office works as a local coordinator and integrator for water governance; it promotes transdisciplinary studies on the local water scarcity and participatory programs. As the office aims at the mutual production of different kind of solution strategies, I classified it as new kind of knowledge production. However, it may also be seen as a new kind of participatory practice.

CONCLUSION

To summarize, this subchapter shows that the institutional system concerning water governance in Petorca is characterized by the market-mode. The field research underlines that cooperation between different actor groups and within the actor groups is low. Analyzing different collaborative projects, it shows that in this institutional framework especially large-scale cooperative projects seem to fail because the lack of willingness to cooperate hinders the development of complex modes of action. Thus, actions are restricted to mainly one-sided adjustments (such as individual technical solutions or migration). Nevertheless, small-scale collaborative projects, so-called social innovations, expand the repertoire of interaction modes. Consequently, they raise the question of whether projects with broader modes of action falsify the assumption that market-based institutions produce restricted modes of action or provide exceptional features that overcome the structures of minimal institutions. An analysis of the underlying mechanisms helps to answer this question. The next chapter explores the mechanisms which deter and/or facilitate modes of action in Petorca and demonstrates how social innovations might overcome the identified barriers.

8.3 Mechanisms of market-based water governance and low cooperation

Chapter 8.1 shows that in the Chilean water governance system (formal and informal) institutions are strongly restricted and characterized by market-style governance. Chapter 8.2 demonstrates a significant lack of cooperation in Petorca. According to Scharpf (2006), in order to conduct solution-capable negotiations, an institutional construct is needed that is more complex than minimal institutions or the classic market style governance as described by Pahl-Wostl (2015). Thus, a connection between institutional construct and cooperation quality is assumed (cf. Scharpf 2006: 92, 197; Ostrom 1999). Consequently, the question to be asked next is what are the explanatory mechanisms behind the restriction, respectively the facilitation of advanced modes of interaction, and are those mechanisms present in Petorca water governance system?

As explained in chapter 5.3, both Mayntz and Scharpf (1995) as well as Ostrom (McGinnes & Ostrom 2014) recommend game-theory to explain those mechanisms. The core mechanisms that influence trust and the cooperative level between actors are communication (as direct as possible); options to sanction opportunistic behavior/ enforcement; inner, experienced heuristics, norms and rules (facilitate or impede cooperation), more specifically we-identity, reputation and fairness; orientation towards reciprocity (positive behavior generates positive behavior and negative behavior generates negative behavior) (Ostrom 2003; cf. Messner 2012). If the institutional system in Petorca is the cause of limited modes of action, these mechanisms might explain the connection between the outcome and the cause. Hence, I used these insights of game theory to find the connecting attributes between the institutional construct and low cooperation in Petorca.

The core mechanisms were taken as a base, to find explanatory connection between the institutional construction and the cooperative level in Petorca. It shows that missing trust is central to all mechanisms. Figure 10 demonstrates a *»scenario of collective action for provision of a public good«* based on Ostrom (2005: 57, cf. chapter 5.3) taking the example of Petorca. It shows a summary of the detected factors that influence the trust level in Petorca's water governance and, consequently, the cooperation level. As the different factors are interconnected, this Figure is a simplified visualization. Several factors increase the heterogeneity of the actor groups, which increases communication gaps and decreases expectations of reciprocity. Further, trust and reputation are negatively influenced by suspicion of conflict of interest and corruption. Interestingly, ecological factors such as the urgent need for water show increasing but also decreasing influence on reciprocity depending on the geographical level.

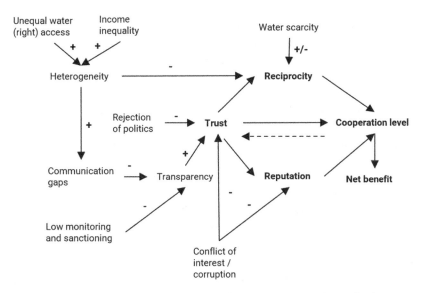

Figure 10: Complex scenario of collective action in Petorca); + means increasing impact, - means decreasing impact. E.g. Income inequality increases heterogeneity. Heterogeneity increases communication gaps and decrease reciprocity. (Own illustration based on Ostrom 2005: 59).

8.3.2 Low level of trust

As shown in Figure 10, trust is a central factor that directly influences co-operation. On a national level, Chile's society is known to have a low interpersonal trust level (3rd lowest in Latin America – after Brazil and Costa Rica, lowest of all OECD countries, Corporación Latinobarómetro 2018). If trust is defined »[...] *as the belief about the probability of reciprocation (Rousseau et al. 1998: 395), or the degree to which people allow themselves to be vulnerable in the absence of external enforcement (Ben-Ner and Putterman 2009).*« (in Messner et al 2016: 53), in Petorca's water governance system, the level of trust seems to be generally low.

During the interviews, members of every actor group emphasize their mistrust of opportunistic behavior by other actor groups, for example, out of fear that they would not divulge important information. Hence, their expectation of reciprocity and their willingness to move to a vulnerable situation by exchanging valuable information or to work together are low. For example, representatives of national authorities suspect that small-scale farmers boycott the JV to benefit from low monitoring (Interview 20). Moreover, representatives of *large-scale agriculture* as well as *local leaders* suspect that *activists* take advantage of »green topics« to pursue political careers (Interviews 1, 23, 29). Members of *civil society* suspect conflict of interest and corruption between *large-scale agriculture* and *governmental authorities* (cf. chapter 7.5.1). Additionally, large-scale projects are rejected because the actors suspect opportunistic behavior. Members of the actor group of *civil society* feel inferior to *large-scale agriculture* or to *governmental authorities*. Therefore, they feel insecure and fear the cost of joining a project together with these actors. For example, representatives of the PGH believe that their program on regularizing water rights for small-scale farmers is slow because farmers do not trust that this process is free of charge. Another actor coming from the group *economy* says that for similar reasons the *Ley de Mono* did not work as people thought it was a trap and they would have to pay a fine afterwards (Interview 25). Further, small-scale farmers believe that *water tanks* or the JV will benefit mostly large-scale farmers (Interviews 12, 14; observation, December 13, 2016). These examples demonstrate that trust is highly linked to reciprocity. In how far reciprocity can be expected to be advantageous to individuals depends on the counterparts' tendency to use reciprocity and on the individuals' capacity to judge in how far the counterpart would use reciprocity in certain situations (Ostrom 2005: 43). In the following paragraphs, I analyze certain mechanisms that could influence the use of reciprocity or the capacity of individuals to judge on reciprocity expectations taking insights of game theory as a base.

8.3.3 Transparency restricted by communication gaps and biases

Several studies show that communication as direct as possible increases cooperation (Ostrom 2005: 29). Ostrom names six reasons why communication increases cooperation in experimental research:

»[…] it [communication] (1) transfers information from those who can figure out an optimal strategy to those who do not fully understand which joint strategy would be optimal, (2) allows the exchange of promises, (3) increases mutual trust and thus affects expectations of others' behavior, (4) adds value to the subjective payoff structure, (5) reinforces prior normative values, and (6) facilitates development of a group identity (Orbell, van de Kragt, and Dawes 1988; Davis and Holt 1993; Ostrom and Walker 1997).« (Ostrom 2005: 33)

Although the first reason seems to be the most important one at first glance, public-goods experiments have shown that the following reasons are equally, if not more important to build trust. Further, it has been shown that face-to-face communication also enhances cooperation (ibid.).

The data presented in chapter 7.5 reveal a lack of direct communication in Petorca water governance. The interviews in this study indicate several communication gaps and biases. Profound gaps divide (a) *large-scale agriculture* and *civil society, (b) governmental authorities* and *civil society,* (c) *governmental authorities* themselves and (d) *civil society* itself.

To a) low-cost local arenas that leave space for direct communication, for example for the purpose of conflict resolution between small- and large-scale farmers, are not present in Petorca (cf. Ostrom 1999: 90). This means that conflicts either go unheard or are carried out in larger cities outside Petorca such as Quillota, Valparaíso or Santiago (cf. Rivera et al. 2016). Intentions to install such local arenas and close the communication gap between small- and large-scale farmers in form of large-scale projects such as the *mesa por el agua* or the *Junta de Vigilancia* were not successfully implemented. Hence, communication between these actor groups is scarce.

To b) qualitative data show that direct interaction between local or regional governmental authorities and civil society takes place on a regular basis (cf. chapters 7.5.1; 7.5.2; 8.2). However, field observations and interviews indicate that this communication tends to be one-directional or biased. One-directional means that, for example, in participatory events about large-scale projects such as technical infrastructure, civil society is informed and may give suggestions or ask question but cannot expect direct answers or a mutual solution finding process (cf. chapter 8.5.1). Biased refers to statements of interview partners from civil society, local authorities and other experts who point out that the number of public institutions is too high, so that several governmental authorities will not coordinate well. This situation leads to overlapping responsibilities and mismatching information (cf. chapters 7.5.1; 7.5.4.3; 7.5.2.1).

The latter leads to c): communication gaps and bias among governmental bodies themselves. Furthermore, these overlapping responsibilities lead to competitive behavior which also restricts information flow between institutions (Interview 29).

To d) Communication gaps have also been identified within the *civil society* group. In Petorca several groups of WUO exist. One could argue that those subgroups provide opportunities for face-to-face communication and for self-organization, for example, for the purpose of cost-sharing and sanctioning (cf. Ostrom 2005: 60). However, information revealed during the interviews shows that all leaders and members of these WUOs complain about the members' lack of participation. Moreover, it was stated that some WUOs compete, such as AC who manage irrigation water and APR who manage potable water (Interview 29, chapter 7.5.1.1). Consequently, direct communication is restricted to those who participate in meetings, and information flow in-between WUOs may be restricted due to competitive interaction orientation.

In view of cooperation incentives in experimental research, it can be said that owing to the communication structures described, it is not likely that the benefits of direct communication are satisfactorily achieved. Findings of optimal joint strategies, exchange of promises, added value, reinforced normative values and facilitation of group identity are limited in the actor groups mentioned. Therefore, trust is not likely to increase and consequently, modes of interaction stay limited. As shown in Figure 10, these communication gaps keep information among actors low. Thus, information trust does not increase, and cooperation remains low as well. A further aggravating factor is that general access to information in this market-style governance framework depends on the actors' financial and educational background (cf. chapter 7.1). This means that especially small-scale farmers, who tend to have low educational background, do not have an overview of the regulative system and its processes. The restricted access and understanding of information, makes spaces of direct-communication even more important. However, communication suffers from gaps and biases. Therefore, transparency of water governance and its actors and actions remains low.

8.3.4 Lack of enforcement

Experimental research shows that options to monitor and sanction behavior may contribute to cooperation (cf. Messner et al. 2013). As Ostrom says: »*Those [individuals] who are initially the least trusting are more willing to contribute to sanctioning systems and respond more to a change in the structure of the game.*« (Ostrom 2005: 37) The findings of this study show that those options are very limited in Petorca. For example, low monitoring of water withdrawal is due to the state's lack of legal and financial regulative power. The national Water Code is criticized by experts and the OECD for lacking environmental and social rights protection. Further, the existing law lacks compliance assurance (cf. chapter 6.4). Two interviewed representatives of the DGA confirm this criticism by emphasizing the current legal and financial conditions that do not allow them to fulfill their work properly (Interviews 22, 35). Further, several interview partners from the actor group *large-scale agriculture* and *governmental bodies and authorities* report that the DGA overshared water rights (cf. chapter 7.5.2 and 7.5.3). By that, the institution undermines its purpose to regulate water rights. Ostrom detected that governance of commons tends to be more successful when »*Monitors, who actively audit CPR conditions and appropriator behavior, are accountable to the appropriators or are the appropriators.*« (Ostrom 1990: 90) In Petorca, the experts of water monitoring are not accountable to the users due to the financial and legal restrictions. Several interview partners underline that this lack of legal and financial resources of the official monitoring body (DGA) is widely known by the users (Interviews 5, 8, 12, 30). Further, the users themselves are not organized to monitor, for example, in form of a JV. Consequently, without appropriate monitoring, graduated sanctions or enforcements are not implementable. Hence, the lack of options to monitor and sanction opportunistic behavior enforces the perception that opportunistic behavior remains unpunished. When the mechanism of enforcement is missing (as in the Petorca case), the concept of reciprocity (positive behavior is answered by positive behavior) cannot be guaranteed as sanctioning rules in case of negative behavior do not work. To make matters worse, members of local authorities and activists report that they have never successful sued in court for illegal water withdrawal (Interviews 8, 16). As a result, they make a negative experience that lowers their expectation of reciprocity in the future. Hence, the information about actors is reduced, the mistrust remains high and the trustworthiness of individuals needs to be assessed via other mechanisms than enforcement (cf. Messner et al. 2013: 15, cf. Figure 10).

8.3.5 Inner, experienced heuristics, norms and rules

As people usually do not possess complete information about all possible gains and costs of certain actions, they draw on learned heuristics, norms and rules (cf. Ostrom 2005: 40). »*Because norms are learned, they vary substantially across individuals, and across time within any particular situation*« (Ostrom 2005: 41). Hence, in how far the low trust level in the Chilean society (lowest of all OECD countries) is connected to past experiences or norms and rules of an individual level would go beyond the scope of this study (cf. OECD 2011). However, game theory offers some indicators to find patterns of systematic interconnections. In the following paragraphs, the norms and rules discovered in the qualitative data of this study are discussed in connection with findings in literature reviews. Specifically, I discuss in how far learned reputation and fairness impact cooperation.

IMPACTS OF HETEROGENEITY

> »Los pilares de la desigualdad en Chile están edificados sobre la base de la apropiación masiva de los bienes comunes«
>
> (The columns of inequality in Chile are built up on the massive appropriation of common goods – translated by author) Rodrigo Mundaca in Bustamante Pizarro 2018).

Large heterogenic groups are less likely to cooperate (cf. Ostrom and Walker 2005). In Petorca actors who directly or indirectly impact the water conflict build a large group – exceeding the provincial level (see chapter 7.4). Further, they differ widely in socio-economic aspects, especially with regard to their financial power and educational background (cf. chapter 7.1). Hence, the heterogeneity of the research case and its impact on cooperation must be considered. Mwangi and Markelova (2014: 462) explain that heterogeinity in groups can be multidimensional (gender, power, wealth and assets, ethnicity, productional system, position in the catchment area etc.) and mostly it increases together with the group size.

Studies of the OECD show that generally low interpersonal trust goes along with low income and high income inequality (OECD 2011). In Petorca, levels of income and income equality are below national average (cf. chapter 7.4). In addition, inequality is reflected in land ownership. Only 7,7 % of all landowners in the Vth region own more than 100 hectares. This is 89,72 % of the existing land. Meanwhile the remaining 80,7 % of farmers own less than 20 hectares which represents only 4,02 % of all land (CNR and Universidad de Concepción 2016: 135). As mentioned in chapter 8.1, such circumstances lead to unequal water access, as actors with higher financial power can obtain more water rights and build better water infrastructure to access water. Low monitoring and sanctioning additionally favors financially powerful actors as their facilities to illegally withdraw water are greater and expectations of punishment are low.[57] The connection between inequality and mistrust is confirmed by the statements of several interview partners. Alfredo and Maria, who both worked with broad parts of Petorca society, indicate that individualistic action is part of the societies culture (Interview 30). Maria believes that individualistic behavior and the focus on personal profit is part of the educational system in Chile and sees this behavior reflected in the low level of cooperation in Petorca. In the following example she explains why small-scale farmers do not cooperate and underlines the connection with inequality:

>»Pero es como que: 'ah no, ellos son grandes, entonces yo no tengo nada que hacer, no puedo pedir nada, y me quedo'. No existen esas capacidades. De por mucho [...] explicarles sus métodos y acciones con sus vecinos y todos. ¿Y cuantas [derechos de agua tienen]? 20? ¿Y cuánto es el que tiene más? 10. Y es el que detiene. Al que no es capaz de enfrentar o de tratar de que las cosas sean más equitativas. 'Ah no, yo tengo 0,5 [derechos de agua] no más y el otro tiene 10.' Lo ven super individualista. No lo ven de una forma asociativa.« (Interview 29)

>(»But it's like: 'ah no, they are big, so I can´t do anything, I can´t ask for anything, and I'll remain like that'. These capacities do not exist. You can explain them methodologies and actions together with their neighbors and everyone as much as you want. And how many [water rights do they have]? 20? And how many obtains the one who has the most? 10. And this is the one who stops. Because they are not able to confront or to try to make things more equal. 'Ah no, I only have 0,5 [water rights] and he has 10.' They take a very individualistic view. They do not consider an associative way.« Translated by author)

Small-scale farmer Iván confirms Maria's observation in so far as he believes to be incapable of changing the situation and repeatedly describes the unequal power relation (cf. Iván 36, 70, chapter 7.5.1.3):

>»[...] o sea y ellos [políticos] mismos son productores específicamente acá en la provincia y entonces contra eso va a ser difícil poder hacer algo. « (Interview 12)

>(»[...], they themselves [politicians] are farmers here in the province; hence, it will be very difficult to be able to do something against this. « Translated by author)

---

57 Financial and legal power of the state to assure norm compliance is low. More details to be found in chapter 8.1.

Representatives of the actor group *large-scale agriculture* have a similar, yet less negative perception of their power. For example, they underline the benefits they have brought to society and stresse that entrepreneurs should have a powerful position as they know how to produce welfare (cf. chapter 7.5.3.1).

Additionally, field observation has shown that the lack of education impedes part of the small-scale farmers to cooperate in different ways. Experts that work with those districts suggest that illiteracy among actors plays a significant role (Interviews 14, 18, 19, 29). Hence, processing information might take longer and could be erroneous, as those actors must rely on face-to-face communication (which lacks). This might be illustrated by an example: In a meeting of the PGH with a WUO an elderly member asked why water rights of his land did not belong to his property. The separation of land and water rights happened in 1981. Apparently, the information needed over three decades to be passed on to this person via face-to-face communication (observation, April 6, 2016). Further, several interview partners report that small-scale farmers were cheated on contracts. For example, they were persuaded to sign papers that they could not read, or they made verbal agreements that are hardly verifiable later (Interviews 2, 15). Such experiences lower the expectation of reciprocity and hence restrict willingness for cooperation. Moreover, they reduce information about possible actions. Exemplifying this with game theory, it means that users do not have access to the rules of the game; therefore, they are cheated or left with at least less than optimal outcome. To sum up, the actors in Petorca's water governance build a highly heterogenic group mainly due to unequal financial and educational backgrounds. As shown in Figure 10, heterogeneity decreases the expectation of reciprocity and expands the communication gaps. Consequently, it influences trust and cooperation level negatively.

LOW WE-IDENTITY
We-identity means that several individuals identify themselves as a group because of one or several common characteristics (cf. Messner et al. 2016). Usually, members of such groups have built up trust easier and tend to use reciprocity so that cooperation is enhanced (cf. Poteete et al. 2010: 44). In Petorca, the heterogeneity and the large size of actor groups hinder a common we-identity for all actors as those factors decrease the collective similarity. One could argue that the plentitude of WUOs could be seen as ingroups. However, because those WUOs suffer from lack of participation and their members tend to show selfish-rational behavior and

prefer individual solutions over cooperation (cf. chapter 7.5.1; chapter 8.2), WUOs cannot be generally considered as ingroups. Hence, an overall collective identity of Petorca' s civil society is lacking.

Nevertheless, the data displayed reveal at least three smaller ingroups in Petorca that show we-identity: large-scale agriculture entrepreneurs, activists and the union of APR. The former group is organized formally as an association (LSAA). It is an association of 55 large-scale farmers who exchange information, cooperate and receive net-benefits. Those net-benefits are diverse, ranging from lobbyism over technological updates to protection of reputational damage. This might be explained by the following examples. The first example stems from the empirical research phase when one of the large-scale agriculture entrepreneurs belonging to the association spontaneously withdrew his interview confirmation. He explained that due to the ongoing water conflict, he was insecure about the negative consequences of such an interview. Instead he used his ingroup insofar as he recommended talking with a lawyer of the LSAA. Although the belonging to an ingroup may favor the insiders, it can increase the risks for those outside. As Messner et al (2016: 55) cite the American developmental and comparative psychologist Michael Tomasello: »*A we-identity allows us to perform acts of generosity and kindness within the ingroup, but as the same time allows us to inflict great damage on those outside it (Tomasello 2009: 99).*« This conclusion is mirrored by the following example: during the interview with a lawyer of the LSAA, he mentions that the group offered the construction of four water tanks in exchange for water rights in a letter to Chile's president Sebastián Piñera in 2012. This letter shows that the ingroup self-organizes and builds up a strategy to overcome negative water scarcity consequences. Moreover, it shows a potential impact on outsiders of that group. If the president had replied positively to the letter, water tanks would have probably been built to benefit all appropriators of Petorca Province. However, the letter could also provoke costs for the group of *civil society* because if water rights had passed to large-scale agricultural entrepreneurs, unequal share of water rights between small- and large-scale appropriators would have raised. According to the lawyer, the president Sebastián Piñera did not answer the letter (Interview 25).

The second ingroup identified in this study is *activists* who are organized in MODATIMA. Although, opinions about strategies vary, members of this group benefit from the solidarity of the we-identity. One may argue that their identity is bound to characteristics such as being politically left or being a small-scale farmer as majority of its members are. However, those characteristics are not exclusive. Members are diverse and consist, for example, of a small-scale farmer who identifies himself as politically right, a local politician and academics. What unifies these individuals is the concern about unfair water management and the motivation to change that. Inside this group, individuals benefit from partnerships and contact with national and international NGOs and lawyers.

However, several interview partners state that such cohesion is lacking in other parts of the actor group *civil society* – especially among small-scale farmers, respectively small-scale water rights holders. Although organized in different kinds of WUOs, actors report that members of such a group do not feel united (Interviews 1, 18). Nevertheless, the communal water provider associations (APRs) in Petorca seem to be an exception to the other WUO. In 2015 APRs established a voluntary union on the community level of Petorca and Cabildo. It was named Unión Comunal de APR Cabildo respective Petorca. The aim of this union is to cope with the ongoing droughts. Inside those unions, APRs exchange information to improve water provision and organize workshops or cooperate on other occasions (such as the *Minga por el agua*[58]).

One could argue that an ingroup exists among the actor group of governmental authorities as well, and certainly there are net-benefits. However, the members of that group acknowledge that competitive behavior prevails, and the discussion shows that the institutional structure does not help to create a we-identity (Interview 29, chapter 7.5.2).

The data show that at least three kinds of smaller groups with a we-identity exist in Petorca. It further indicates that the ingroups do not connect actors of different groups but stay either in *civil society*, or *large-scale farmers*. Yet, even closed inner groups do not form strong we-identities, especially among members of *civil society* and among members of *governmental authorities* a we-identity was not found.

---

58   For more information about *Minga por el agua* please consider chapter 8.2.

REJECTION OF POLITICS AND WATER GOVERNANCE

This lack of we-identity is further connected to the rejection of politics. Several interview partners express rejection of politics or report about rejection of politics in society and connect this to the lack of social organization respectively cooperation (Interviews 16, 29, 30). The qualitative data indicate that these individuals experienced uncooperative behavior in the recent history. One could argue that this perception is rooted in the dictatorship of Agosto Pinochet. For example, a small-scale farmer says that during the time of the dictatorship, even small group meetings were dangerous. She argues that this is the main reason why cooperation and organization in small groups is still perceived as something forbidden, bad or rebellious (Interview 36). As demonstrated in chapter 7.1, surveys show that Chileans still have low trust in political parties and the government (Corporación Latinobarómetro 2018). Voter turnout in presidential elections remains one of the lowest worldwide (Ríos 2017) and surveys attest an above average rejection of democracy (24% according to Pew Research Center 2017). Hence, the rejection of politics goes along with disenchantment of democracy and is universally aimed at all political directions. The surveys indicate that after 30 years of democracy in Chile, society is not convinced about the benefits of this type of regime. Scholars like the Chilean sociologists Barozet and Espinoza (2016) have explored that the rejection of politics fostered during democracy. They argue that this happened because parties did not respond to the needs of the developing society that turned more and more towards individualistic values. During democracy, water governance did not experience any drastic regulative changes (cf. chapter 7.4). The constitution was maintained; the neoliberal premise of water privatization continued and was extended by privatizing the public water providers. As explained in chapter 8.1, the state´s power has remained limited since 1981 and the reform in 2005 did not have any direct impact on *civil society.* Consequently, I claim that democracy has not brought any significant improvement in water governance and politics in so far, so the society has not been able to benefit from political participation. These power structures are widely perceived by society resulting in the observed cognitive perceptions and individualistic behavior as detected in this study via analysis of interaction orientation (see chapter 8.2).

Moreover, interviewees state that people reject organizing and cooperating when they feel that it could be a political act. In addition to individualistic behavior, they find a second reason for not cooperating: Individuals are afraid of the cost of being identified as a political left or right oriented person. Thus, they tend to avoid certain organizational activities that might have something to do with politics (Interviews 8, 16, 30, 36). The actor groups of *activists*, *local leaders* and *economy* are an exception of this political apathy. Most of the *activists* and *local leaders* identify with the political left and demand higher regulations in water governance (cf. chapter 7.5.1). Their extraordinary interest in politics can be explained by their role in society that requires cooperation and an understanding of water governance politics. The political values of *large-scale agriculture* show a tendency towards neoliberal and right-wing politics as they support the market-style governance (cf. chapter 7.5.3). The exception to rejection of politics of the actor group *economy* could be related to education and financial resources as studies show that people with a higher education tend to reject democracy less, and voter turnout is higher in areas with wealthier people (Ríos 2017).

The qualitative data show that political apathy in Petorca reflects the general Chilean rejection of politics. This rejection in water governance might be explained by the institutional system that is characterized by low governmental power and extensive complexity. Hence, actors perceive that politics, governmental bodies and institutions cannot assure reciprocity and do not see benefits in participating. Moreover, political apathy in Petorca is associated with the concern for personal reputation, as actors prefer not to be considered political. Actors fear the cost of reputational damage. Consequently, trust to participate in (social) organization respectively cooperation remains low. The significance of reputation is further discussed in the following paragraphs along with the importance of fair reciprocity.

PEOPLE ALWAYS TALK ABOUT... REPUTATION
The *reputation* of individual actors and actor groups is strongly connected to the learned heuristics, norms and values in water governance. As Messner et al (2016: 54) say: »[...] *we seek information about others' past performance to try to guess how they will behave in future.*« If there are no institutions that assure transparency or norm compliance, actors must rely on rumors or experience to estimate if the other is trustworthy. Thus, the reputation of an actor or an actor group may influence cooperation. In Petorca various actor groups have a negative reputation that prevents other actor groups from cooperating with them as cooperation is closely linked to the expectations of reciprocity and fairness. To illustrate this interrelation, the reputation of different actor groups and its impact on cooperation should be clarified. For example, in Petorca the actor group of large-scale farmers is described as »unscrupulous« by various other actors, especially

by civil society (Interviews 1, 3, 4, 7), but also by local governmental authorities (Interview 15) and a large-scale farmer (Interview 23). The criticism mainly refers to overexploitation of water rights and its impact on environment and society. Likewise, activists suffer from reputational damage. Different actors from *governmental authorities* as well as *civil society* consider them as violent and refer mainly to protest marches (Interviews 7, 11, 17). In addition, some interviewees share experiences of rejection in the interaction with others due to their being perceived of as an activist (cf. chapter 7.5.1.2). Further, governmental authorities have the reputation of being corrupt or in conflict of interest with large-scale agriculture entrepreneurs. This reputation is spread by way of the media as well as personal information. For example, activists refer to political scandals that were circulated through the national media to explain why governmental bodies should not be trusted (e.g. Interview 5). Moreover, members of the actor group of *civil society* and *large-scale agriculture* refer to rumors and experiences on corruption. Corruption is often used as the explanation and justification for their rejection of governmental authorities (cf. chapter 7.5.1). These examples show that the reputation of an actor or an actor group influences the willingness of others to cooperate (see Figure 10). In the case of Petorca several actor groups suffer of reputational damage, which contributes to the explanation why cooperation is low. The discussion of the qualitative data shows that especially powerful actor groups, such as large-scale farmers, who are powerful in terms of financial resources and governmental bodies that are powerful in terms of political decisive power are perceived negatively by the numerical largest actor group civil society. The following paragraphs explore the role of power and its connection with reciprocity and fairness.

»SO, DO YOU THINK THAT THIS IS FAIR?«[59]

The accusations of conflict of interest and corruption show that in order to cooperate, actors do not only need the expectation of reciprocity, but also reciprocity needs to be perceived as fair (cf. Messner et al 2016: 54). External authorities can have an impact on rule changes as they can influence a fair implementation and compliance of rules – especially in strongly heterogenic groups (Ostrom 1999: 274-275). However, as explained in chapter 8.1, the power of governmental bodies to assure law compliance is low

---

59   Translated by author. Original statement of small-scale farmer: »Entonces, ¿tú crees que eso es justo?« (Interview 12)

in Petorca water governance. Hence, the expectations of fair reciprocity are low as well. Further, closeness to public officials may make one actor or one actor group more likely to change the rules for their own profit than others. Particularly in Chile, mayors are known to have great influence on the implementation of local projects (Espinoza 2013: 402-403). In a corrupted system, this can even lead to a situation in which rule changing is no more possible due to the loss of confidence to cooperate (cf. Ostrom 1999: 200-201). In Petorca, respondents criticize the corruption and conflict of interest and the unequal power structures and unfair laws. Thirteen interview partners suspect corruption or conflict of interest. Most of them belong to the actor group of *civil society,* such as *local leaders, activists* and *citizens*. However, also representatives of local *governmental authorities'* state that corruption is a problem in Chilean water governance. While the representatives of the municipalities only state that corruption might be happening, other local officials assure that corruption is indeed a problem in Petorca. Furthermore, one member of the actor group *large-scale agriculture* assures that corruption undermines the water governance's functionality and says that a state's official tried to bribe him (Interview 23).

Many of the interview partners perceive that rich people dominate decision- and rulemaking in water governance (cf. Interviews 5, 16, 28). For example, members of *civil society* believe that large-scale agriculture entrepreneurs influence the DGA (Interviews 4-6) and the law (Interviews 1, 5, 12). The issue of the former minister of the interior Perez Yoma (2008-2012, 1. Mandate of resident Michelle Bachelet) is often taken as an example for that. Interviewess tell that the owner of the large agricultural company, located in the province, was repeatedly accused of illegal water withdrawal and commissioned the ex-director of the DGA as his lawyer. Because of that closeness to politics, and specifically to the DGA, considerable number of small-scale farmers and activists suspect corruption, which harms the trust in the governmental system. Moreover, these suspicions are nurtured by national political scandals such as corruption allegations against president's Bachelet family members (2015) and the toilet paper cartel (2015) (cf. chapter 7.1).

In general, several members from almost all actor groups state that they perceive the Chilean water governance or related systems as unfair (22 interviewed persons). Various interview partners see the unequal power structures reflected in the water regulation, which according to them favors large-scale economic players. Expectedly, the actor group of economy is an exception. However, one large-scale agriculture entrepreneur supports this perception (Interview 23). While one of the interview partners from the actor group of *civil society* accepts these structures, and believes they pave the way to progress (Interview 9), others perceive the water management system as unfair (cf. chapter 7.5.1):

»Entonces yo creo que acá hay un tema de distribución. Él que tiene más poder está perjudicando a los más chicos, o él que tiene más dinero, porque como te vuelvo a repetir, o sea si fuese sequia todos estaríamos complicados, es una cuestión lógica esa yo creo. Pero eso no se ve. Solamente él que está más afectado es el más chico.« (Interview 12)

(»Hence, I believe that it's about a distributional issue. The one who has more power is harming the smallest ones, or the one who has more money, because, I will repeat, well, if this would be a drought, all of us would be in trouble, this is logical, I think. But you don't see that. Only the smallest ones are more affected.« Translated by author)

Moreover, members from civil society, governmental authorities of regional and national level state that water regulations are unfair in so far as they follow a win-maximizing strategy that supports mainly big economic players (cf. Interviews 17, 22). As a consequence, they sense that vulnerable members of society are left behind. For example, a representative of the national DGA reports that WUOs are not democratic and dominated by large-scale agricultural entrepreneurs (Interview 22). Additionally, an actor of the group *economy* claims that the win-maximizing strategiy of the regulative system does not respect the limits of the ecosystem as he reports:

»Yo estaba en ese marco y ese marco no me funcionó. ¿Por qué? Porque no se cumplió con la ley básica que el agua es escaza. Y si el estado sigue otorgando derechos y derechos y derechos y resulta que al final terminan todos muertos. Y más encima entre medio entregó plata a todos porque para hicieron proyectos para que desarrollaran, le dieron subsidios, le regalaron plata.« (Interview 23)

(»I worked within this system and it didn't work for me. Why? Because it didn't follow the basic low of scarce water. And when the state continues to give away right after right after right, in the end everyone will end up dead. And on top of everything the state gave money to everyone so that they can do their project and develop. They gave them subsidies; they gave away money.« Translated by author)

Furthermore, the rejection of large-scale projects above described reveals the importance of lack of fairness (cf. chapter 8.2). Although, small scale farmers could profit from a *Junta de Vigilancia*, as it could improve monitoring and sanctioning opportunities, and as a representative of CORFO says: *»no decision could be made that does not follow the law«* (translated by author, Interview 20), they reject the project. They feel that the decision making is not fair because a large-scale water right owner's vote counts more than a small-scale water right owner's vote. Similarly, large-scale water tanks are rejected because small scale farmers fear that their share would not be as high as that of others. Another example comes from small-scale farmer Iván who perceives the process of a water infrastructure construction as unfair:

> »Porque es igual que estuviéramos en una mesa y yo los invito a almorzar y somos cinco personas y yo a cuatro les sirvo el plato o les sirvo el pan y al otro lo dejo mirando, considerando que estamos en la misma mesa ¿Cómo se va a sentir ese quinto?« (Interview 12)

> (»It´s like as, if we would be all at the same table and I invite everyone for lunch and we are five persons and I will serve the meal to four persons or I will serve bread and I will ignore the last one, considering that we are all at the same table: How will this fifth person feel?« translated by author)

These examples demonstrate that the power structure, dominated by financial wealth and information, is not only subtly inherent in the governmental bodies and structure of market-style water governance, as explored in chapter 7.1. Moreover, it is perceived widely by different actor groups. Consequently, individuals do not only limit their actions to the wider range of possible action modes but also to the expected outcomes. For example, several interview partners indicate that they would not report water robbery, nor take any other actions which would be legally possible as they feel powerless inside the existing structures (cf. Interviews 6, 12, 14, 28, 29). The discussion indicates that in Petorca's water governance institutions norms and rules are not strong enough to guarantee the actors fair reciprocity. Therefore, cooperativeness among those actors remains low.

8.3.6 Ambiguous impact of ecological factors

Since this study applies a socio-ecological framework to analyze water governance in Petorca, it requires a discussion about ecological factors. According to Ostrom, rule changes depend on the resource condition and its measurability or predictability (Ostrom 1999: 269-270). She claims that in order to organize, it is important that actors find incentives in the resource scarcity in so far that a completely exhausted or an abundant resource will not motivate to self-organize (Ostrom 2009: 420). In Petorca, two opposing impacts of water scarcity are observable. Some interview partners indicate that holders of water rights from the Petorca District would cooperate more than those from other districts as there is generally more water in the Petorca River basin (cf. Henríquez et al. 2016). They advocate Ostrom's assumption as according to them; when there is no water people see no point in organizing, for example, to construct water tanks or to clean channels (Interviews 1, 15, 29). The *Junta de Vigilancia* is an example of the divergent cooperation levels in relation to water scarcity. While efforts were made to form such a WUO in both basins, only in Petorca River the process of building a *Junta de Vigilancia* advanced. Hence, the existence of water seems to correlate with the expectations of reciprocity during actions in water management. On the other hand, experts fear that a natural drought alleviation would take out political pressure to change water governance structures because with the end of the drought not only people withdraw from participation in water right organization, but also political decisions will be slowed down:

> »[..], por eso es que en estas reformas no están apurados porque si vienen siete años de lluvia capaz que estas reformas al código de agua ni siquiera sirvan, porque [ellos dirán]: 'no. ¿para qué prioridades de uso si hay tanta agua?'.« (Interview 28)

> (»[..] and that is why the reform is taken time because if there will come seven years of rain, most probably these water code reform will be useless, because [they will say]: 'No. Why do we need priority in water use if there is so much water?'.« translated by author)

Indeed, a representative of the Ministry for Agriculture indicates that there is no need to change rules because the farmers adapt naturally and recently precipitation has increased (Interview 17).

Hence, according to the interview partners, on a local scale people expect fewer benefits from cooperation under dry conditions. On a larger scale, experts indicate that agreements might be reached faster in times of drought as the need for reciprocity is higher and fairness factors are not as significant. However, the rise of small-scale project also indicates a pressure for cooperation on a smaller scale during drought. All small-scale projects found in this study emerged in respond to water scarcity impacts. On the whole, the impact of ecological factors towards cooperation remains an interesting question for further investigation.

8.3.7 Conection factors: institutional entropy and inequality

The mechanisms between low level of cooperation and institutional framework show a mutual connection between two factors: the multitude of weak governmental authorities and the socio-economic inequality. The following discussion explores these factors in connection to the filtered mechanisms and concludes chapter 8.3.

Chile's governmental bodies and regulations concerning water governance show a retroactive development that adapts to current needs post-factum. For example, the DGA was installed in 1969 as pressure rose for the need of a specific water regulating institute. As irrigation water forms the most important water use, the CNR was established as a department of the Ministry of Agriculture in 1975 (Peña 2018: 199). This was the beginning of the development of a multitude of governmental bodies dealing with water issues. Chile's water governance regulation is based on the Water Code of 1981, which was created following neoliberal paradigms of a weak state, aiming at enhancing economic growth. Consequently, it limits the state's power and assigns it to private water right holders. Governmental bodies and the set of institutions around it were designed to function in a free market setting following the assumption that the balance of powers is regulated by competing actors, demand and supply. However, these conditions were not fulfilled as the socio-economic system that influences water governance and its different interrelated units turned out to be more complex. In a response to that multiple governmental bodies were installed afterwards. Literature offers several numbers of Chilean (governmental) bodies that directly or indirectly deal with water issues reaching from 43, counted by the World Bank (Banco Mundial 2013), to over 100 (Retamal et al. 2013). In addition to those governmental bodies, several WUOs exist. The regulative structure of those WUOs follows similar patterns of minimal institutions, and its power structures are aligned with economic wealth (cf. chapter 7.4).

In its environmental performance review of 2005, the OECD mainly praises the Chilean water management because it has been able to drastically increase water provision and sanitation while installing full cost recovery (OECD 2005). While the OECD connects this progress with privatization, the organization also mentions that active trade remains uncommon (ibid.). The organization's recommendations mainly concentrate on ecological needs such as implementing an integrated watershed approach or improving industrial and housing sewages treatments. Further, they emphasize the improvement of information and fundamental knowledge. In 2005 the Water Code was reformed to address social and ecological needs. The passive character of the regulative system can be seen in the long discussion of Water Code reforms that took approximately 15 years for the first Water Code Reform in 2005 and more than five years for the currently discussed Water Code Reform. The reform of 2005 intended to enhance the regulatory function of the state and to assure higher environmental standards. However, these changes did not profoundly alter the institutional landscape of water governance, nor did they overcome divergent power structures. The reform aimed at incremental improvement of the regulatory system to assure the functionality of the water market (cf. Bauer 2015: 263). For example, the reform introduced taxes-for-non-use, which means that water rights which are not used for economic use or domestic provision are fined. This new rule was supposed to prevent water right holders from speculating. All in all, the reform has been criticized for being too weak. For example, the OECD/ECLAC (2016: 75) criticizes that taxes for non-use are too low to incentivize water trade. Further, the reform did not include constitutional reforms, its changes apply to water rights only that have not been handed out yet, and it does not imply a prioritization of water for human use (cf. Bauer 2015: 265). As a result, the reform did not alter the architecture of governmental authorities and minimal institutions. Humberto Peña, ex-director of the DGA between 1994-2006, defended the introduction of the reform in 2005 by stating that it would bring an equilibrium between social and economic needs and between public and private interests. However, when he evaluated the same reform eight years later, he revised his conclusion and found several flaws in the reform of not being capable to protect environmental and social needs (cf. Bauer 2015). Regarding the institutional system and the multitude of governmental authorities he states:

>>The emergence of new problems has led to fragmented institutional development in a bid to solve problems sector by sector, creating multiple bodies and mechanisms each with a limited and partial perspective way in which hydrological processes occur.<< (Peña 2018: 201)

Gopalakrishnan's (2005) describes similar processes of functionality loss and has coined the term institutional entropy. It describes the incremental rise of institutional dysfunction and underlines the historical dependency on the one hand, and complexity of ongoing changing institutional environment on the other hand. He states:

»In many countries, resource management institutions initially designed to perform at top efficiency have failed to make adequate and appropriate adjustments to allow for the many changes - political, technological, legal, and cultural - that inevitably accompany the passage of time.« (Gopalakrishnan 2005:4)

Such institutional entropy is reflected in the increasing number of water conflicts in Chile. The diagnosis of Chilean water management of the World Bank in 2011 (Banco Mundial 2011)[60], points to several flaws that were detected in this study as well, and offers recommendations which address especially the socio-ecological needs and the strengthening of actors (governmental authorities as well as WUOs). Also, the second OECD environmental performance review on Chile from 2016 makes recommendations addressing the dysfunction of Chilean water governance, which were not dealt with in 2005. The review highlights social needs such as prioritizing water for human consumption and sanitation, enhanced implementation of round tables as water conflict resolution strategies, and strategies that strengthen monitoring, sanctioning and transparency (OECD/ECLAC 2016: 26). Several of these recommendations were adopted by the current water reform discussion, for example the implementation of an ecological basin and the prioritization of water for human use (cf. chapter 7.4).

The increasing development of conflicts of water governance is also reflected in Petorca's water governance. As water scarcity and unequal water access have worsened during the last seven years of drought, institutional flaws that might have existed earlier emerged because of the socio-ecological pressure. This study defines water governance as sustainable if it integrates strategies for long-term environmental and social resilience by using the benefits of participation and cooperation. Due to its retroactive character, the institutional construct in Chile does not provide long-term strategies. In Petorca, instead of finding strategies for mitigating or adapting to climate change impacts, short-term ad hoc solution strategies like water provision by trucks may last for years. Furthermore, new technologies such as desalinization plants or the use of groundwater are lacking in regulative frameworks and public acceptance (cf. Peña 2018).

---

60 The diagnose was requested by the first government of Michele Bachelet (2006-2010). The publication of further in-depth studies was planned and announced by Sebastian Piñera for the second phase of 2011. However, they never have been published. (cf. Bauer 2015: 287).

This lack of long-term and sustainable strategies is connected to the obstacle that the institutional framework and its multitude of governmental authorities do not (sufficiently) address problems of inequality. Consequently, they do not incentivize the benefits of participation and cooperation. Considering the cooperation restricting mechanisms filtered in this study, it becomes clear that both institutional entropy and inequality go hand in hand in each of those mechanisms. Financial and educational inequality increases the heterogeneity of the group. These inequalities are promoted in the regulative system as it gives the decisive power to the water holder according to their number of water rights. Furthermore, the multitude of governmental authorities, overlapping responsibilities and competing structures makes it difficult for actors to participate - especially for those with a low educational background.

Despite the intention to balance power structures by the reform in 2005, water governance did not manage to address social needs. Learned heuristics, such as the rejection towards politics, the suspicions of conflict of interest and corruption or other opportunistic behavior that impacts reputation loss, decrease the expectation of a fair reciprocity. Moreover, a lack of opportunities for monitoring and sanctioning water use was detected, especially for less powerful actors who do not find institutions - neither in terms of reliable rules and values, nor in terms of responsible governmental bodies or authorities. All those mechanisms dissolve trust. Further, the institutional system does not overcome the communication gaps and bias which are strongly impacted by educational inequalities. Therefore, information about actors and actions, in other words transparency, in the group remains rather low which also restrains trust for cooperation and hinders the development of net benefits.

To sum up, the path towards sustainable water governance in Petorca contains several pitfalls that consist of different mechanisms that are rooted in a market-style institutional framework. These mechanisms are triggered by institutions that are not powerful enough to address problems of socio-economic inequalities. On the contrary, the multiplicity of governmental bodies and authorities with its overlapping responsibilities and weak power seems to prevent formal institutions from being applicable and understandable for a great part of the involved actors (cf. Beckenkamp 2014: 54). Therefore, the existing institutions do not offer possibilities to gain trust in order to enlarge the possible modes of interaction or do so insufficiently. In other words: The traditional institutions of water governance in Petorca neither provide mechanisms that build enough trust, nor do they allow sufficient assessment of the costs and risks of cooperation.

Consequently, the institutions do not provide additional modes of interaction which would overcome the market-based governance style. In the following, I take the outcome outlined here and compare it to the social innovations referred to in chapter 7.2 to discuss if and how they overcome the cooperation restraining mechanisms.

8.4 Leverage points for enhanced cooperation

Several mechanisms have been identified that impede increase of trust and reciprocity and therefore decrease cooperation level (cf. chapter 8.3). However, small-scale projects were found that seem to overcome these mechanisms and show more diverse action modes (see chapter 8.2.3). To obtain information on the strategies of those social innovations, qualitative data from interviews with members of those initiatives were analyzed. Since the information was incomplete, additional interviews with founders or members of the initiatives *petorQuinoa, Banco de Semilla* and *MODATIMA* were conducted in a third field research stay in 2017 (Interviews 31-34). Different mechanisms (M_2) were filtered that helped the social innovations to overcome the hindering mechanisms (M_1). The mechanisms are displayed in Figure 11, which is based on Ostrom's »*Complex Scenario of Collective Action for Provision of a Public Good*« (Ostrom 2005: 57). It shows an overview of the trust and cooperation decreasing mechanisms filtered in chapter 8.3. Further, it displays trust and cooperation increasing strategies of social innovation in red color and how they interplay within the scenario. In short, it exemplifies how strategies of social innovations feed into the existing structure and create levers to increase trust, reciprocity and cooperation. The following paragraphs describe these levers in more detail.

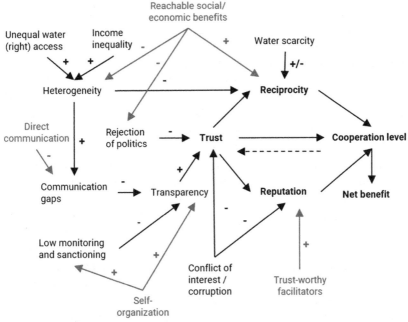

Figure 11: Impact of social innovations (colored in red) on complex scenario of collective action in Petorca. + means increasing impact and - means decreasing impact, e.g. heterogeneity increases communication gaps but decreases reciprocity. (Own illustration based on Ostrom 2005: 59)

8.4.1 Fostering transparency with the help of trust-worthy facilitators

The transparency of Petorcas' water policy is low, on the one hand due to limited access to and limited understanding of information, and on the other hand due to communication gaps and prejudices, in particular towards *civil society, government agencies*, and between *government agencies and civil society* (cf. chapter 8.3). Several social innovations target this issue. By using direct communication strategies and benefiting of trust-worthy facilitators, they overcome communication biases and gaps, increase information bases, and therefore increase trust (see Figure 11: direct communication). On those visits, they educate water right holders about the Water Code and the current modification processes.

Further, they offer their services of counselling about water rights and agricultural management issues. After each event they offer tea and cookies to create an opportunity for informal discussion. Their communication strategy targets two different mechanisms. First, it gives the opportunity to close the communication gap between PGH and civil society, and between WUO members themselves. Second, it offers information that might help to close knowledge gaps caused by the low educational background. This repeated face-to-face communication supports trust building over time. In addition, the PGH benefits from staff members who have made a name for themselves as trust-worthy persons, for example, one staff member has worked with small-scale farmers in Petorca for several years and another staff member has been a known member of the community as she was born in Petorca and teaches at a local college. Several WUO members emphasize that they would trust these people because they have already met them in other contexts (observation, December 12, 2016). Hence, those trust-worthy people facilitate cooperation.

The other projects presented here follow similar face-to-face communication strategies towards *civil society* that involve trust-worthy people as facilitators. For example, the project *Banco de Semillas* was promoted by a teacher that approached small-scale farmers by visiting their homes. Further, the project was supported by students who benefit from the trust inside their families. They told their parents or other relatives about the *Banco de Semillas* and used the trust inside their family to make the small-scale farmers of their families cooperate.

MODATIMA was mainly built by a group of local small-scale farmers. They have organized events and talked on a face-to-face basis to neighbors, farmers, local leading person to convince them of joining the movement. The central figures of that group are known people in the province. Pablo, for example, worked for approximately a decade in agriculture counseling in Petorca and building trusting relationships with several members of the agricultural area. Other central figures are in leading positions as well, such as WUOs presidents and municipality counselors. Hence, the core members are in positions where they can approach a large part of the community and other *local leaders* while facing apparently low barr   iers. Having gained the reputations to be trust-worthy individuals facilitates their attempts to motivate cooperation.

In continuation of the list, the association *PetorQuinoa* also was launched by a local leading figure. Luis Soto is known for his campaigns for the mayor's elections and his work with *MODATIMA*. By using his local network, he closed the gap between *governmental bodies* and *civil society*. This can be illustrated by an example. Members of *PetorQuinoa* report that after having been rejected by several governmental bodies, CORFO offered to fund Luis' idea of building a network of small-scale farmers under one condition that he should submit a list of at least 100 interested farmers. According to the interview partners, Samuel convinced more than 100 farmers to join the project out of which *PetorQuinoa* emerged (Interview 31).

A similar inter-connecting strategy was identified in the project *Oficina de Asuntos Hidricos (OAH)*. The person responsible for the office, Irene, succeeded in connecting *civil society* and *governmental bodies* by founding the *OHA* in the Petorca District. Irene has an academic background and is not originally from the Province. However, she knows several WUOs from her previous work at the *PGH* and strengthened her proximity to the community when she moved to the Petorca District. She then used her academic and local network to implement the *OAH* as a sub-office of the municipal department for community development. According to her, she keeps contact especially to the *Unión de APR* and acts as a link between water right holders and *governmental bodies*. Her daily communication tool with the community is a group on the smart phone app »WhatsApp«. Further, she can be reached in her office on weekdays (Interviews 14, 32). The project *Minga por el Agua* can be taken as an example of how Irene works as a facilitator. Irene and Ana took over leading roles in communicating the project to other WUOs as well as *governmental bodies* to encourage them to work together. Ana has already been known as a leading person in her community because of her local activism. Further, her leading role as a president of a WUO shows that she is considered a trust-worthy person. However, she mentions that Irene plays a central role in motivating and connecting different actors (Interview 3). Hence, both, Irene and Ana, work as facilitators for enhanced cooperation.

These examples show how the social innovations bridge communication gaps and improve access to information. They thus promote transparency, build trust and achieve advanced modes of action (cooperation, cf. Figure 11).

8.4.2 Overcoming heterogeneity and rejection of politics by highlighting social and economic benefits

Heterogeneity and rejection of politics are identified mechanisms that keep cooperation level low in Petorca's water governance (cf. chapter 8.3). Especially actors of the group *civil society* expect either unfair reciprocity or fear high risks in cooperation, for example losing their reputation. The projects mentioned here solve this problem by finding common interests or needs that are not perceived as political in the first place. Instead of focusing on the cognitive perception of the situation's causality, for example the question if climate change or monoculture cause water scarcity, they focus on motivational aspects, the actors' identity and interests such as basic preference for self-preservation (cf. Scharpf 2006: 116). By emphasizing achievable social or economic needs, they address several cooperation restricting mechanisms (M_1) as shown in Figure 11. They circumvent the obstacle of politic rejection and develop subgroups that keep the group size low and diminish heterogeneity. Thus, they increase expectation of reciprocity.

PetorQuinoa, for example, emerged as a response to the question of how to stop economic loss of small-scale farmers in times of water scarcity. The cooperative offers an answer to that problem and allows farmers to stay in Petorca and in the farming business. The expected benefits and costs of this project are not abstract but can be calculated. Divergent opinions, such as cognitive aspects on how and why water scarcity occurs in Petorca, are overcome by addressing motivational aspects such as the wish to remain independent as a small-scale farmer. The same motivation to support small-scale farmers is claimed by the *Banco de Semillas.* Additionally, they refer to the protection of traditional seeds. Both reasons target at economic, respectively socio-environmental, benefits that are closely linked to the identity and interest of their project partners. At the same time, those benefits are achievable and calculable which lowers cooperation risks. *Minga por el agua* might have the most calculable and achievable benefit of the social innovations presented. The installation of a water pipe allows participators to calculate how much effort they would like to put into cooperating (working hours, tools etc.) so that risks remain low. The expected direct benefits will primarily reach those living in the project area. However, participating actors not living in that area could expect indirect benefits such as an improved reputation and greater expectation of reciprocity when repeating a similar action situation. The *PGH* also places socio-economic benefits at the center of its work. They claim that cooperating with them will help small-scale farmers to find alternative cultivating strategies to cope with water scarcity. Furthermore, they offer to assist in legal processes to clarify the water right situation in order to gain access to governmental water programs (such as JV). The *PGH* offers their services for free, so once actors trust the employees at *PGH,* they can calculate the costs and risks of cooperating.

The activists's movement MODATIMA works differently. Their demand of »recuperar el agua/ reclaim water« is rather abstract. However, they also find a common motivation by seeking for fair water access and distribution. Their members differ in their cognitive perception of the situation because of their different professional backgrounds and political opinions. Nevertheless, they agree to cooperate in the expectation of a rather abstract reciprocity, such as the benefits of belonging to an ingroup or the personal motivation of doing »the right thing«. The letter indicates that this cooperation is strongly connected to their motivational aspects such as their identity. This can be illustrated by some examples. The activists describe that they work for a better future (Interviews 5, 6, 8). They seem to identify with the character of a heroic fighter (cf. chapter 7.5.1.2). This identification seems to be a core element of the group's social cohesion.

One could argue that the *OAH* tries to put *MODATIMAs* view of an alternative water governance into practice. The *OAHs* aim is to strengthen WUOs, especially APRs, and establish a transparent public water governance. Although this goal is rather abstract, on an operational level the *OAH* concentrates on social and economic benefits, for example, by building a common hub for agriculture tools or organizing informational events. Cooperating actors expect to benefit from the shared information and the connection to various governmental bodies.

As shown in the examples above, highlighting social or economic benefits of social innovations increases trust, reciprocity, and, consequently, network benefits. Scholars of critical institutionalism recognize a similar factor. They stress that introducing strategies that turn the community into a `service provider` is more beneficial to the communities than the more commonly discussed factor of ownership (cf. Whaley and Cleaver 2017: 60). Additionally, the social innovations shown here avoid conflicts that are triggered by divergent cognitive aspects such as the normatively charged discussion about water right property. This does not mean that those topics are not discussed internally, but it lowers the barriers to entry into cooperation.

8.4.3 Establishing information and fairness by self-organization and institutional bricolage

As described above, institutions that assure transparency of information and norm compliance (monitoring and sanctioning) and fair reciprocity seem to be lacking. However, appropriate monitoring of the common-pool resource could increase the actors' information and therefore, increase the trust levels among them. If the monitoring is organized by the users themselves, according to Ostrom, the increased amount of information held by the actors will help them to make decision about whether they will profit of a rule change or not (Ostrom 1999: 96-98). Hence, the development of institutions from within the group can foster trust (cf. Beckenkamp 2014: 55). The social innovations presented here show self-organization that allows actors to install rules and norms which offer transparency and are considered as fair. As displayed in Figure 11, self-organization enhances information about actions. These self-determined institutions are exemplified in the following.

For example, the cooperative *petorQuinoa* agreed on making decisions by votes. Unlike traditional WUO voting systems, each member has only one vote regardless of his or her share of the cooperative (Interview 31). That way, small-scale farmers can expect a rather fair reciprocity as the voting system aims at balancing the power structures. Additionally, they allow members to sell products privately if necessary:

> »La idea es que los productores nos pasan todas sus quinuas a la cooperativa, pero no existe ninguna obligación a respeto a eso. Porque entendemos también que es muy difícil para un productor. De repente necesita luca y si puede vender quinua el, que lo venda. Pero queremos avanzar que al final la cooperativa haga mejor el negocio que ellos por sí solo. Entonces solo nos van a pasar la quinua. Porque es mejor para ellos.« (Interview 31)

> (»The idea is that the farmers pass all their quinoa to the cooperative, but there is no obligation to do this. Because we understand that it is hard for a farmer. Sometimes he may need money and can sell the quinoa, so he should sell it. But we want to progress so that in the end farmers make a better business inside the cooperative than by themselves. Then, they will give us the quinoa because it's better for them.« Translated by the author.)

By this, they set their own rules for cooperation. Small-scale farmers retain their independence by being able to calculate cost and benefits each time they want to contribute to the cooperative. In this way *PetorQuinoa* keeps the cooperation barriers low and makes it possible to gradually increase trust in repetitive action situations. Thus, they enable learning circles that are based on empowerment and self-responsibility and aim at increasing reciprocity.

Other social innovations also provide examples of self-built institutions. The *Banco de Semillas* has decided to install a monitoring system. They keep track of all incoming and outgoing seeds and supervise different irrigation techniques used to build and share knowledge that can deliver the highest levels of productivity (Interview 33). *Minga por el agua* was organized by a local community. The community members approached trust-worthy outsiders such as WUOs and *local leaders*. They decided how to reward or thank the helpers (e.g. offering meals). Self-organization allowed for transparency about the work process and the expected reciprocity of their actions. Further, they gained independence from *governmental bodies* which are not considered as trust-worthy. The *Oficina de Asuntos Hídricos* emerged with the support of the local *Unión de APR* which comprises approx. nine thousand members of *civil society*. The central objective of the *OAH* responds directly to the lack of institutions as it aims at establishing public organized water governance. A central task for reaching this goal is to install new institutions that provide transparency. In this way actors increase their information on the action situation and will be empowered to make decisions based on greater knowledge about possible risks and benefits.

MODATIMA is a self-organized environmentalist movement. They have decided to have an organizational structure that includes a president, a secretary and a PR manager among others. At the same time, they do not want to be considered an organization but a movement. They have agreed on a common position that criticizes the current water governance. The actions and decisions made by the members are communicated within the ingroup. However, not everyone agrees with each strategy, and members are free to oppose. For example, when voting for the Water Code Reform several members decided to support the reform at the congress while others kept their distance and criticized that the reform did not go far enough. Since these installed rules are self-made, members may rely on those rules with a higher expectation of reciprocity.

The *PGH* cannot be defined as a self-organized project because it has hierarchical organizational structures. The main objectives are set by its main organization *Pontificia Universidad Católica de Valparaíso* (PUCV) and must be indicated in annual reports. Project relevant decisions must be agreed upon in advance with the office director. In addressing small-scale farmers, the *PGH* pursues a strategy that aims at cooperating with farmers who have certain promising characteristics. However, the *PGH* still allows for space for creating own institutions. For example, in addition to proposing their own ideas, the *PGH* workers repeatedly call for proposals of new cultivating and marketing strategies. In face-to-face counseling meetings PGH workers and small-scale farmers mutually develop business plans. In this way, small-scale farmers have a greater opportunity to increase information that helps them to calculate risks and benefit of cooperation and trust towards the *PGH* workers increases.

8.4.4 Remaining challenges of social innovations

This study detected three main strategies that increase cooperation in social innovations. However, not all projects use the same mechanisms in the same way, and all projects face different pitfalls. The following displays the main obstacles named by the members of social innovations and discusses it the light of the market-style water governance and its impact on interaction.

To understand the context in which the social innovations described here have emerged, it is not enough to highlight their success strategies, but it is also necessary to explore their difficulties. Interview partners from different social innovations report about similar problems. Two challenges are outstandingly named: the lack of motivation of actors to cooperate and bureaucratic barriers. Furthermore, it shows that the social innovations only succeed to incentivize cooperation between one or a small number of actor groups (cf. chapter 8.2). None of them includes the actor group of large-scale agricultural entrepreneurs. The following paragraphs describe these problems in detail.

Firstly, despite their success, members of social innovations report that there is still a barrier to motivate individuals to participate. They state that this is mainly because of their mistrust (Interviews 14, 32), and their fear of high personal risks. The problem of restricted cooperation and its connection to an expectation of (fair) reciprocity was explored in chapter 8.3. This may be illustrated by some examples from the experience of social innovation. In the case of the *Minga por el Agua*, a small group of neighbors did not participate. They thought that they would not benefit from the project because they believed that the water installation would never reach their homes (Interview 3). Hence, they did not expect any reciprocity. Members of *MODATIMA* report that citizens are afraid of losing their jobs at large-scale agricultural companies if they participate in the activist's activities. An outstanding example for those personal risks is the agronomist

Rodrigo Mundaca. He and other members claim to have received threatening calls and other intimidation attempts (Amnestía Internacional 2018). *PetorcQuinoa,* on the other hand, is not confronted with a lack of interest and therefore has no problems recruiting members. However, several members of *PetorQuinoa* report that inside the cooperative participation is lacking and leading task are carried out by a few members only (Interviews 12, 31). Therefore, Fernando claims that there is a need for social skills on how to organize and motivate members (Interview 31).

Secondly, the interviewees report about struggles with administration and bureaucracy that make it hard to understand and conduct legal processes. Especially, members of *PetorQuinoa* state that administrative processes slow down the progress of their cooperative. Fernando stresses that in addition to the usual difficulties of bureaucracy, the concept of a cooperative is new to many public and private organizations, which complicates the administrative processes (Interview 31). The civil servant of the *Oficina de Asuntos Hídricos* stresses that the bureaucracy consumes a lot of time and money. Hence, socio-economic less powerful actors have difficulties to handle the bureaucracy. She states that in order to deal with the divergent types of actors (private and public), she urgently needs administrative help (Interviews 14, 32).

The project of *Banco de Semillas* showed additional challenges which are more related to the planning of the project itself. This means according to them, that they have problems with the commercialization of their product in terms of finding consumers and that they have technical problems with irrigation. However, they managed to find solutions for both issues (Interview 33).

Thirdly, with regard to the social innovations in the light of sustainable water governance, another problem should be mentioned. The main actors of such social innovations belong to only one or two actor groups. Hence, they enhance cooperation and trust, among these groups but not in between different actor groups. In particular, no social innovation involves the socio-economically most powerful group of agri-business entrepreneurs. Considering that those actors probably have the biggest ecological impact on the region, sustainable water governance without addressing those actors is not achievable.

CONCLUSIONS

This chapter filters mechanisms that help social innovations in Petorca to incentivize cooperation. The three main mechanisms are (a) applying direct communication and increasing information especially by trust-worthy local facilitators, (b) highlighting social or economic benefits (motivational aspects such as identity and interest) to avoid conflict over cognitive aspects and (c) applying institutional bricolage by self-organizing. All of these help to increase the expectation of (fair) reciprocity, to improve information about reputation and increase trust. In this way they facilitate cooperation. However, it has been shown that the actors involved in those social innovations claim that mistrust and low motivation to cooperate remain challenging. Furthermore, deterring bureaucratic barriers were detected. In addition, none of those social innovations cooperates with the actor group of large-scale agricultural entrepreneurs. Based on the discussion of the current institutional framework, including its cooperation deterring and facilitating mechanisms, the next chapter provides recommendations for a water governance architecture that aims at sustainability.

8.5   From here to where? Towards polycentric water governance on river basin scale

The aim of this study is to evaluate the solution capacity of the current water governance system in Petorca (Scharpf 1997: 90) and to determine whether a shift towards another welfare production is possible and which restructuring such a change would imply. The guiding paradigm is to reach sustainable water governance that aims at adaptive strategies towards long-term environmental and social resilience and uses the (known) strategies to implement an integrated management system. This study shows that the capacity of Petorca's water governance of reaching cooperation is low. Consequently, large-scale solution strategies that aim at alleviating the water conflict have not been implemented successfully. Scharpf emphasizes that the institutional construct needed to succeed solution-capable cooperation is more complex than market-style governance. This research confirms his argument by showing that modes of actions are rather limited (cf. Scharpf 2006: 92, 197; Ostrom 1999) and sustainable solution cannot be reached (as shown in chapter 8.1 to 8.4) if formal and informal regularization inherent to the institutional system is as low as in the Petorca water governance system. However, social innovations were detected that overcome the institutional context and lead to a higher cooperation level. Still, the following question remains: which institutional construct is possible in Petorca that allows sustainable water governance and what are the steps to achieve it?

The following paragraphs provide an answer to this question by suggesting an adaptive governance approach that recognizes and tackles complexity. In this way, it offers directions towards sustainable water governance for Petorca based on the insights of this study. Considering the complexity and context bound characteristics that would make any »manual« for an ideal outcome questionable, I focus on designing the process towards sustainable water governance rather than giving guidelines (cf. Pahl-Wostl 2015: 21).

In the context of current political and academic discussions and the empirical research of this study, I argue that Chilean water governance must move away from neoliberal paradigms and minimal institutions towards an integrative governance model that aims at balancing power structures and enhance cooperation. Leaning on Ostroms (2005: 283) insights on polycentricity and Pahl-Wostls (2015: 8, 22, 44) emphasis on self-organization combined with purposeful design, I argue that a *polycentric water governance* focusing *on a river basin scale* management structure may provide the most suitable and solution-capable institutional framework in Petorca and even Chile.

Moving away from market-based principles, I suggest that a shift towards a more network-oriented governance mode may be the most achievable improvement. Such a governance mode can emphasize self-organization and, in that way, allow cooperation facilitating mechanisms that are nurtured by social innovations.[61] However, two characteristics of the current system need to be taken into consideration: (1) the already fragmented and weak formal and informal institutional structure and (2) the high level of heterogeneity, especially with regard to the socio-economic inequality among the actors. In view of this, a purely network-oriented governance mode of self-organization might not overcome pitfalls of unequal power structures (cf. Pahl-Wostl 2015: 89). Therefore, I emphasize a mixture of governance modes in so far as the network-structure of self-organization should be surrounded by governmental bodies and authorities to ensure rule-compliance and transparency (cf. Ostrom 2005: 283). In that sense, it combines bottom-up and top-down strategies, an approach also called meta-governance (Ostrom 2005: 59, Pahl-Wostl

---

61   For example, direct communication, institutional bricolage, etc. (see chapter 8.4)

2015: 95). It is important to note that meta-governance not only signifies a mixture of different governance strategies (bottom-up and top-down), but also a constant learning process. I am referring to Pahl-Wostl (2015: 96) who defines: »*meta-governance as a reflexive process of societal learning to develop, to evaluate and to adapt governance approaches with the purpose of addressing complex societal challenges'*«

The deciding factor in any case is the manageability of this institutional system and the recognition of the problems of an already fragmented structure. Hence, the institutional system must clarify responsibilities, legal structures and communication to avoid increased fragmentation. In this regard, it must strengthen institutions and introduce or allow the building of new ones. It must especially strengthen the role of the community in form of WUOs (financing, self-organization, democratic structures) and the monitoring and sanctioning role of the state in form of the DGA (financial and legal power). Due to extensive controversy about the topic of water privatization versus nationalization, the suggested approach focuses more on community governance than on ownership, and more on decision-making processes than on the outcome to avoid exceeding transaction costs (cf. Bauer 2015, cf. Pahl-Wostl 2015). Nevertheless, I argue that water for human consumption must be prioritized to secure a minimum of distributional justice and that management must recognize the ecological boundaries of a minimal river flow.

The following Figure 12 exemplifies the governance architecture of the suggested polycentric water governance on a river basin scale taking the example of the Petorca River. It basically concentrates on the structure of politics (actors and political processes). It is important to note that this study recognizes indirectly correlating systems in water governance, as it argues that sustainable governance must include those indirectly correlating systems as well. Therefore, it might appear inconsistent that the following image highlights actors more directly concerned with water management. This is to simplify the visualization and does not intend to exclude actors and organization of correlating systems (such as agriculture, mining, NGO´s etc.).

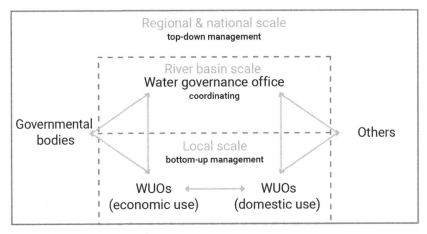

Figure 12: Polycentric water governance on river basin scale. (Own illustration)

In the following the suggested governance structure is described in more detail. To structure the paragraphs, I will first demonstrate polycentric water governance according to its polity and politics (referring to the institutional framework, actor constellation and processes) and later regarding its policy (content and implementation of rules) (cf. Treib et al. 2007).

8.5.1 Politics and polity suggestions

Based on the outcome of this study and current research, I recommend a so-called meta-governance that combines bottom-up and top-down strategies that are coordinated by a polycentric governance office (cf. Pahl-Wostl 2015: 95). First, this text focuses on a design for bottom-up. Then, it displays the combination with top-down strategies.

PARTICIPATORY, RESPONSIVE, EQUITABLE AND INCLUSIVE SELF-ORGANIZATION OF WUOS COORDINATED BY WATER GOVERNANCE OFFICE
The need for cooperation in water governance is common ground in current research (cf. UNESCO 2006). Petorca is determined by market-oriented principles and insufficient traditional mechanisms. However, social innovations emerging from within the community have shown that they overcome these traditional mechanisms (cf. chapter 8.4). Nevertheless, the question remains to which degree these new creative strategies can pave the way towards a more solution-capable, sustainable water governance. In that sense, this dissertation decodes mechanisms that facilitate

cooperation between actors on a local level. Hereby, it highlights context-specific indicators that might lead to more sustainable water governance. First, it shows that space for self-organization is recommendable. Hence, I stress the importance of space for institutional bricolage, i.e. ways to establish common agreed institutions, for example to monitorand sanction water use. Second, direct communication – especially among *governmental bodies* and *civil society* – as well as inner communication of those two actor groups is highlighted. Third, it stresses that social or economic aims should be in the foreground. These aims need to be more concrete and achievable than abstract. Furthermore, people identified as trust-worthy by the community within the local network help to facilitate in all three issues. I suggest that the design of polycentric water governance in Petorca should facilitate these mechanisms by promoting self-organized bottom-up structures that give more decisive power to community members and WUOs. Regarding community governance, Whaley and Cleaver ask a crucial question about functionality: »[...] *what is being implied for the role of communities and about their capacity to perform these roles?*« (Whaley and Cleaver 2017:60).

To answer this question, it may be useful to consider the multiple levels of analysis by Ostrom (1999, cf. chapter 7.4). According to her, the suggested polycentric governance provides a maximum of self-organization and allows WUO and community members to impact all three levels – operational rules, collective-choice rules and constitutional-choice rules. Operational rules determine, monitor, apply and enforce day-today decisions, while collective choice rules determine »*who is eligible to be a participant and the specific rules to be used in changing operational rules*« (Ostrom 2005: 58). Lastly, constitutional rules decide who can participate in a regulative framework and how it is built. Despite the precarious work situation faced by WUOs, management tasks are already being adopted as defined by Ostroms (1999: 65).[62] That is why I suggest an institutional framework on a local level that works in a network-oriented governance mode. By giving space for self-organizing, institutional bricolage might take place. That means that new institutions are build or existing ones are combined. In this way trust and reciprocity will gradually increase.

However, the current institutional framework analyzed in chapter 8.3 and the social innovations analyzed in chapter 8.4 demonstrate several pit-

---

62   E.g. informal status, lack of economic resources, low educational background (cf. chapter 7.4).

falls that need to be considered in the design of polycentric governance system. In relation to the former, this study shows that different WUOs, and members of the same WUO, are in competing positions. At the same time, they are characterized by high heterogeneity, mostly because of their unequal socio-economic backgrounds. Hence, empowering those structures might not necessarily lead to higher welfare but foster unequal distribution. As Messner (1997: 205) explains: *»[...] networks are able to reach the Kaldor optimum only when they are not guided by competitive, or indeed hostile, action orientations vis-à-vis other network participants (in the sense of »getting the best of the others«)«.* Therefore, it is crucial that the institutional framework on the local level aims at equalizing power structures by seeking more consensus-oriented decision processes (e.g. one person – one vote).

Regarding the latter, social innovations show three main difficulties that should be recognized when trying to implement their cooperation facilitating mechanisms. First, they do not involve different actor groups but remain consisting of one or two actors, and exclude, in particular, the socio-economically most powerful actor group: large-scale agriculture entrepreneurs. Second, they face organizational obstacles mainly rooted in bureaucratic issues and in a lack of trust among actors (cf. chapter 8.2 and 8.4).

Dealing with the highly fragmented institutional landscape, including its diverse actors and cross-sectional interests in water and the interrelations of its socio-ecological system, could easily exceed the capacity of the local community. Therefore, I suggest that those local WUO management processes should be coordinated on a river-basin scale. This coordination should guarantee transparency by coordinating communication among the WUOs and with actors outside the local context (e.g. governmental authorities, NGOs, universities) and by providing information and knowledge management. The purpose of this coordination in addition to the organizational benefits is to allow for learning circles that might help to increase trust and reciprocity. As Ostrom says:

»Because polycentric systems have overlapping units, information about what has worked well in one setting can be transmitted to others who may try it out in their settings. Associations of local resource governance units can be encouraged to speed up the exchange of information about relevant local conditions and bout policy experiments that have proved particularly successful. And, when small systems fail, there are larger systems to call upon – and vice versa.« (Ostrom 2005: 283).

In other words, a coordinating body provides space for learning by repetitive experience of (fair) cooperation. In particular, such a coordination body is important to coordinate the integration of actors in systems that directly and indirectly influence water governance, to identify knowledge and communication gaps and bias, and finally to coordinate education and learning. Since these tasks should improve the processes in both directions, top-down and bottom-up, it becomes obvious that such a coordinating body requires sufficient staff and finances. The *Oficina de Asuntos Hídricos* in the Petorca District has already taken over a part of the coordination tasks mentioned here and could be taken as an already existing, yet improvable, example for good practice. In how far it coordinates communication, information and knowledge is shown in the Box 2.

Example of Oficina de Asuntos Hídricos as a coordinating body

An example for such a coordinating body is the *Oficina de Asuntos Hídricos* in Petorca which is funded by the municipality and listed as a social innovation in this study. The office works as a facilitator for the self-organization of the WUOs – especially concentrating on APRs. Hereby, it focuses on communication and knowledge management (cf. chapter 8.2).

Communication: the office coordinates the communication between the members of the individual WUOs by using social media and organizing meetings and workshops. Further, it facilitates the informational flow by coordinating the communication of the WUOs with corresponding governmental authorities, NGOs, universities or research centers. Besides contacting the different partners, it also helps the WUO members to express their interest.

Information and knowledge management: the purpose of the office is to facilitate self-organization and educate WUO members and other citizens. It organizes workshops that inform the members about the rights of water users and workshops that educate them in water quality management. Further, the office director underlines the necessity of transparency and information. Therefore, the office aims at installing an online and physical library that allows community members to access relevant information. (Interviews 14, 32)

Box 2: Description of coordinating body taking the example of Oficina de Asuntos Hídricos

Some may argue that a similar bottom-up structure has already been provided by the possibility of building a *Junta de Vigilancia*[63] (cf. Vergara Blanco 2014). In the Petorca Province the building process of a *Junta de Vigilancia* has not been successful so far. In view of the results of this study, I argue that it has failed to succeed because it does not overcome the cooperation deterring mechanisms discovered in chapter 8.3. In particular, it does not allow for democratic decision making to overcome inequality. Moreover, it does not have a facilitating and coordinating body as suggested here.

FOSTERING EFFECTIVENESS, EFFICIENCY, AND RULE COMPLIANCE BY TOP-DOWN MANAGEMENT

This study shows that the institutional framework is characterized by minimal institutions that promote the consequences of the socio-economic heterogeneity of its actors and limit cooperation due to problems of reputation, lack of monitoring and sanctioning as well as communication and information gaps and bias. As illustrated in Figure 12, I suggest self-organized network-oriented polycentric water governance on a river basin scale which is surrounded by a broader institutional framework with a hierarchical orientation. I intentionally use the word orientation as it is intended to underline that diverse combination and hybrid forms should be possible. To avoid misunderstandings, it should be repeated that this is not an instruction guide but rather a suggestion for general directions, based on this study. This proposed broader institutional framework with a hierarchical orientation is represented by governmental bodies and authorities which are primarily responsible for overcoming minimal institutions by ensuring rule compliance.

Following Ostrom (1999), they monitor compliance of constitutional-choice and/or collective-choice rules. As the overall institutional system has shown to be very fragmented, it is important that problems of communication gaps and biases as well as overlapping responsibilities among the states authorities are tackled. Therefore, it is necessary that those institutions have enough legal and financial background to work effectively and well-organized coordination to work efficiently. This can be illustrated by examples: (1) the possibility of unannounced inspections would allow higher hit rates for the DGA, and (2) the water governance office on river basin scale could help coordinating the communication of those govern-

---

63   For information about Junta de Vigilancia please consider chapter 7.4

mental bodies with the community. Hereby, it can function as a feedback to states authorities and help to detect and solve problems of communication gaps and overlapping responsibilities. Hence, a more transparent, effective and efficient management of rule compliance might allow actors to increase reliable information, experience reciprocity and therefore increase trust. However, the existence of organizations working hierarchically on monitoring and sanctioning does not exclude monitoring and sanctioning mechanisms on the operational-choice level. It rather promotes rule compliance and ensures transparency.

8.5.2 Policy suggestions

REGULATORY FRAMEWORK AS A BASE FOR ACCOUNTABLE AND TRANS-
PARENT WATER GOVERNANCE
This study defines sustainable water governance as an integrated water governance that recognizes the interplay of political, social, economic and administrative systems concerning water, that considers correlating systems outside of the water sector and aims at adaptive strategies for long-term environmental and social resilience. Consequently, such complexity-embracing paradigms need to be reflected in a regulatory framework that aims at sectoral integration and copes with uncertainties and risks. In that sense, the regulatory framework must be flexible enough to adjust to context bound socio-ecological local conditions. However, the current regulative framework in Chile has been criticized mostly for being too weak (OECD/ECLAC 2016; Belmar et al. 2010, Retamal et al. 2013). Currently proposed reforms have been debated for several years. These debates emphasize inter alia the prioritization of domestic use, the end of infinity of property rights, and the acknowledgement of a minimal ecological flow. Considering the complexity of rules inherent to a governance system Ostrom states:

> »No one can undertake a *complete* analysis of all of the potential rules that they might use and analytically determine which set of rules will be optimal for the outcomes they value in a particular ecological, economic, social, and political setting. [...] Further, since ecological, economic, social and political settings are always changing over time, no specific set of rules will produce the same distribution of benefits and costs over time« (Ostrom 2005: 255)

Accordingly, it is not within this study's scope to give detailed recommendations on policy. However, regarding the demonstrated socio-economic inequalities of the water users, the socio-ecological consequences and the high level of mistrust, I argue that a regulative framework is needed that ensured the following three demands of the current discussed Water Code reform: the prioritization of water for domestic use, the end of infinity of property rights and the introduction of a minimal ecological flow. Such a regulatory framework should ensure accountability in both directions: WUO self-organization strategies and hierarchical-orientated governance of governmental authorities. Furthermore, the regulatory framework prepares the base for maintaining water management within the ecological limits, and it must be applied at the various levels of governance (cf. Ostrom 2005: 258).

CONCLUSION

Finally, I underline that the polycentric architecture for a sustainable water governance in Petorca is intended to be a strategy that highlights meta-governance in the sense that at a combination of effective governance styles is sought (cf. Pahl-Wostl 2015: 95). The individual sub-functions of water governance such as policy framing, knowledge generation, or conflict resolution are neither characterized as bottom-up, nor top-down modes only. Rather, it a purposeful design, based on the present study that aims at encouraging cooperation by giving space for self-organization combined with trust facilitating institutions. The development of such sustainable water governance is considered to be time-intensive as it emphasizes an incremental process of change that is nurtured by learning circles (cf. Pahl-Wostl 2015) The following argument by Ostrom serves as a holistic conclusion:

> »Polycentric systems are themselves complex, adaptive systems without one central authority dominating all of the others. Thus, no guarantee exists that such systems will find the combination of rules at diverse levels that are optimal for any particular environment. In fact, one should expect that all governance systems will be operating at less than optimal levels given the immense difficulty of fine-tuning any complex, multitiered system.« (Ostrom 2005: 284)

Against this background the combination of bottom-up and top-down orientations suggested in this study should create openess and underlines the importance of polycentric coordination bodies that facilitate the learning processes towards sustainable water governance.

Part 5: Conclusions

Starting from the global need for sustainable water governance to cope with and/or mitigate water crises, this study took the water conflict in Petorca (Chile) as a case to investigate how the institutional framework impacts water governance and which may be the steps towards sustainability (cf. Part 1). As ecological and social systems influence each other, I chose to use a socio-ecological system perspective to analyze the institutions and actors' behavior (cf. Part 2). The results show that the market style water governance in Petorca restricts trust and cooperation. However, social innovations were detected that overcome those restricting mechanisms. Afterwards, I transformed the lessons learned from this study into a reasonable design approach towards sustainability. I recommend polycentric water governance on a river basin scale (cf. Parts 3 and 4) that uses the cooperation strengthening mechanisms of social innovations from bottom up and installs reliable institutions from top-down. To reduce the problems of the fragmented institutional system, this process should be coordinated by a polycentric water governance office. The following chapters conclude this study by highlighting the empirical findings, theoretical and practical implications, limitations and recommendations for future research.

9. Empirical findings

The empirical findings of this study range from a broad analysis of secondary, mostly quantitative data about the social and ecological setting, through the governance system to self-collected qualitative data on actors and action orientations. The quantitative data suggest that social and financial inequalities in the Petorca Province are above national average and expose the water scarcity impacts. Those impacts mainly affect agriculture, which is the most important economic activity, plus it is the most water demanding sector in Petorca. In particular, it causes excessive job losses for small and medium scale farmers (cf. chapter 7.1). In the Vth region, those small and medium scale farmers are the largest group (80.7 %), account for only 4.02 % of cultivated land (CNR and Universidad de Concepcion 2016: 332). It should be emphasized that the largest part of the land (90 %) is owned by large-scale farmers (ibid.). Further, the extensive research demonstrates that water scarcity is related to a long lasting severe drought (cf. Garreaud et al. 2017). Although dry periods are natural in central Chile, data indicate that this drought is independent from the El Niño Southern Oscillation (ENSO) and is part of the anthropologic climate change impacts (cf. chapter 7.2). Both river basins in the province (La Ligua River basin and Petorca River basin) have been declared to be dried out since 2004 respectively 1997 (DGA 2018). What aggravates the situation is that water rights in Petorca have been granted on too large a scale. Further, it was detected that a considerable number of water rights in Petorca have not been clarified, i.e. they have not been correctly registered. This situation causes problems for water right holders to organize in WUOs. It shows that in addition to WUOs, governmental bodies, which directly or indirectly influence water management, are diverse. This is related to the historical continuity of water governance in Chile. In line with paradigms of neoliberalism, current water regulation starts historically with the introduction of the Water Code and the constitution of 1981. Both aim at a reduced state power and a free market. The core of these regulations has remained valid until today (cf. Bauer 2015). However, negative social and environmental developments were observed. In response to that, reforms have been made, and governmental bodies have been created to prevent such negative outcomes. This reactive manner, however, does not change the neoliberal paradigms of water governance as, for example, water right privatization is promoted by the constitution. Moreover, water governance is criticized for lacking transparency and rule compliance assurance (cf. chapter 7.4).

This setting is reflected in the actors' perceptions and interaction orientations. Civil society mostly feels left behind and perceives the system as unfair. In particular small-scale farmers and activists accuse large-scale agriculture of leaving the communities without water and blame the water regulation and governmental bodies of favoring those companies. Further, they criticize governmental bodies for their overlapping responsibility, and a lack of transparency and functionality (cf. chapter 7.5.1). Governmental authorities present a more ambiguous picture. While some actors of this group agree with civil society inter alia by stressing that the regulative system encourages inequality and favors large-scale agriculture, others claim that civil society profits from the lack of monitoring and sanctioning by the regulative system. Moreover, the interviews expose that some actors in leading positions do not believe in anthropologic climate change (cf. chapter 7.5.2). Representatives of large-scale agriculture and the water providing company report about economic loss but are mostly content with the water regulation. Instead of political changes they insist on technological solutions. One actor of the group of large-scale agriculture, however, complains heavily about the insufficient monitoring and sanctioning of water use and about corruption in water management. He claims that this situation, caused by governmental authorities and large-scale farmers, has led to an ecological crisis and to the economic loss of the entire province (cf. chapter 7.5.3).

The institutional setting is organized as market-style water governance based on limited rules, norms and values in the sense of a reduced state and price-controlled motives and power structures (cf. chapter 8.1). Additionally, representatives of all actor groups worry about the opportunistic behavior of others and low cooperation. This concern is also shown in their orientation of interaction which is mainly selfish-rational or competitive and lays the foundation for the low cooperative level. The latter is not only perceived by the actors but also reflected in the organizational shortcomings of the various water users' organizations and the failings of large-scale projects requiring improved modes of interaction (cf. chapter 8.2). The mechanisms behind market-based water governance causing low cooperation are related to mistrust and low expectations of reciprocity. The most prominent mechanisms are (1) lack of communication and communication bias, (2) lack of enforcement, (3) social inequalities leading to a lack of transparency as well as (5) conflict of interest respectively corruption allegations and (6) rejection of politics that cause reputational damage and reduce the expectations of reciprocity (cf. chapter 8.3). Nevertheless, in this sea of individualistic behavior, small islands of improved cooperation can be found. They exist mainly on a local scale and involve one or two actor groups only. The social innovations apply varying leverage points to overcome the institutional framework. The most prominent mechanisms are (1) direct communication facilitated by trust-worthy persons; (2) prioritization of achievable and concrete goals that address motivational aspects more than cognitive aspects to overcome rejection of politics and mistrust; and finally (3) self-organization that allows for institutional bricolage, i.e. the development of new institutions that help to improve transparency (cf. chapter 8.5).

Based on the empirical findings and the discussion part of this study, I recommend polycentric water governance. I argue that this is a design that allows benefits of cooperation to grow by promoting water user organizations. It provides operational space for those organizations to make use of the leverage points described above. Further, it strengthens rule compliance by encouraging top-down monitoring and sanctioning of water use.

Since quantitative and qualitative data indicate high institutional fragmentation and socio-economic inequality, I argue that these processes need to be coordinated by a water governance office on river basin scale that focuses on communication and knowledge management to ensure transparency and learning circles towards sustainability (cf. chapter 8.5).

10. Theoretical and practical significance of this study

This dissertation aims at refining and generating of hypothesis by case study research. It contributes to studies of socio-ecological system and institutionalism in by applying a synthesis of two well-known analytical frameworks (social-ecological system framework (SESF) by Ostrom (2007, 2009) and actor-centered institutionalism (AI) by Mayntz and Scharpf (1995)). In this way, my research aims at testing their variables and empirical indicators and at discovering alternative causal interrelations (cf. chapter 4).

By considering insights of social innovations research and critical institutionalism, it proves that the selected analytical framework is flexible enough to adapt to the current research debate. Further, this study introduces a novel approach to social innovation studies that puts this rather new strain of thought into an elaborate theoretical frame.

This study provides context-bound data that underline Mayntz and Scharpf's theoretical assumption about minimal institutions causing restricted modes of action (cf. chapters 8.1 to 8.3). However, it also proves that this assumption is not of absolute validity. Exceptions show that, on a local scale, actors can overcome the institutional frame by elaborating their own norms, rules and values (cf. chapter 8.4). By merging the SESF and AI, both approaches broaden their variables. The former benefits from a systematic analysis of the action situation by specifying the indicator »mental models« in concrete terms of cognitive and motivational (interests and identity) aspects and orientation of interaction. Further, the AI helps to analyze and categorize the variables collected according to SESF in the category of governance systems. The indicators for describing the action situation offered by SESF partly repeat and overlap with the sub-tiers of further categories (such as social, economic and political settings, resource systems, governance systems, resource units). Meanwhile, the AI offers an alternative way to describe the action situation, based on the information gathered via indicators of the first tiers from SESF. Conversely, SESF contributes to the AI by placing the analysis in a more contextual perspective that acknowledges the influences of the ecological system on the actors' behavior and vice-versa. Further, two indicators were developed in the discussion of the results to analyze cooperative quality. Those are neither explicitly stated by the SESF, nor directly mentioned in findings of game theory: collaborative projects and rejection of politics. The former can be detected via other indicators of the SESF. However, it was useful for this study to take the findings of social innovation to categorize such projects and analyze their organizational structure and the circumstances under which they emerge. While the SESF has sub-tiers like »self-organizing activities« or »deliberation processes«, collaborative projects as introduced in the discussion of this study help to point to cooperation characteristics of these activities that are not necessarily self-organizing or deliberating. Rejection of politics was detected as an indicator that influences trust and cooperation level. In this study, the interview analysis revealed that various actors report that people do not cooperate because they fear the cost of reputational loss and have low expectation of reciprocity due to their perception of politics and of people they relate to politics. To analyze the reasons behind that rejection, the quantitative and qualitative data gathered via the indicators from SESF were useful. Although I do not claim to have fully investigated these reasons, I emphasize that rejection of politics, depending on the case at hand, may be useful for the behavior analysis and therefore I recommend adding it to the list of variables.

Further, this dissertation contributes to the overall discussion of sustainable water governance by a detailed case study that stresses current arguments for an integrated design. Firstly, it indicates that market style governance is not recommendable for achieving sustainability. Secondly, it highlights the importance of integrating system thinking and bottom-up approaches.

These contributions have potentially practical implications. Firstly, this study reveals the gap between international resolutions signed by the Chilean government to ensure the right to water and sanitation, and the local reality of citizens living in rural areas. Secondly, this dissertation provides science-based recommendations for an institutional design that helps to rebuild trust, enhance cooperation and move towards sustainability: polycentric water governance on river basin scale. This purposeful design is not only formed based on the insights of water governance research as such, but also on the real political situation in Petorca as an emblematic case of water conflict in Chile. Consequently, it is not abstract but suggests achievable structures. Similar concepts have already been discussed in the Petorca Province. In 2016, the governor of the Petorca Province suggested a water office that should pool and react to concerns about water from civil society as well as coordinate responsibilities of different governmental bodies in order to decrease bureaucracy and higher transparency (observation, November 30, 2016). In the Petorca District, such an office called *Oficina de Asuntos Hidricos* was built as a department of the municipality. This dissertation shows that such a coordination body forms a leverage point towards enhanced self-organization of the water user groups (bottom-up management). However, to ensure rule compliance, I argue that stronger governmental bodies are needed that build up reliable institutions and higher expectations of reciprocity (cf. chapter 8.5). Thirdly, this dissertation confirms that the paradigms of neoliberalism, such as privatization and reduced state, affect trust and reciprocity and consequently cooperation negatively. The case of Petorca shows that a change of paradigms is needed to reach sustainability ecologically and socially. In order to reinstall trust into politics and governmental bodies, corruption allegations and conflicts of interests need to be clarified, respectively avoided. The current power structures, as shown in this study, undermine the state and favor economic actors. This study does not provide detailed policy recommendations as this would exceed its scientific scope. Rather than providing policy instruction, this study points to patterns for a more sustainable water governance. This means, politics should strive to rebalance power structures instead of intensifying them. To give a concrete example, in 2018, the government of Sebastián Piñera aimed at privatizing the National Water Directive (DGA), the governmental body in charge for monitoring water right use (el periodista 2018; El siglo 2018). This study exposes scientific arguments about why such decision may impair trust and reciprocity in Chilean water governance even further. Furthermore, two concrete policy recommendations are made in this dissertation: in order to ensure right to water and an environmentally healthy environment, it recommends prioritization of water for human use and the implementation of an ecological basin as currently discussed in the Water Code reform.

11. Limitations of this study and recommendations for future research

This study takes a socio-ecological perspective in water governance and recognizes that society and environment are closely related and interdependent. It aims at a holistic and contextual approach to explore institutional patterns and actors' behavior. My educational and scientific background and the basis of this study are social sciences, respectively, political sciences. For the aims of this dissertation, I cooperated with scientists and experts from engineering and natural sciences in Chile and Germany. They predominantly helped to find and understand data of the ecological system. Nevertheless, the capacity to include the flows that go in and out of the ecological systems was limited and the focus was placed on the social system. In the following paragraphs I demonstrate some examples of the limitations that, at the same time, provide incentives for future research. This study indicates that cooperation increases on a smaller scale as water scarcity diminishes. On a larger scale, however, scarcity alleviation may block political attempts at institutional change. Consequently, I recommend further research on the connection between both systems. Additionally, this dissertation highlights social innovations as a trust building engine for improved cooperation. As these strategies respond to water scarcity in a new creative manner and aim at higher quality of life; they are supposed to target sustainability. Therefore, it is interesting for further research to establish how these social innovations influence the ecological system. Further, the OECD/ECLAC (2016) criticizes water quality monitoring, particularly, in rural areas in Chile. During this study strong concerns about water quality in Petorca were raised by members of civil community. How to improve those monitoring systems in rural areas technologically and in terms of management is another interesting research question to be addressed by an interdisciplinary team of scholars.

Additional limitations of this study were a lack of capacities and time for a thorough analysis of financial flows such as taxes or municipal spending on water trucks. An in-depth study of those may highlight the economic impacts of the scarcity not only on directly affected individuals such as small-scale farmers but also on the government, and it may encourage changes towards sustainability. The same applies to further studies on the interrelations of international players and on corruption allegations and conflicts of interest.

Due to the context-based nature of case-oriented studies, the developed analytical approach and results must be adjusted respectively and tested in other settings in order to be generalized. This thesis offers an analytical framework that filters mechanisms of the institutional framework and actors' behavior in a socio-ecological system perspective. Following this perspective will contribute to a better understanding of the urgent question in real life politics and science: how to shape the transition towards sustainable water governance. Finally, I recommend working in an interdisciplinary team to capture insights into the interrelations between social and ecological system.

References

Agrawal, Arun (2001) 'Common Property Institutions and Sustainable Governance of Resources', World Development, 29(10): 1649-1672.

Aitken, Douglas; Rivera, Diego; Godoy-Faúndez, Alex; Holzapfel, Eduardo (2016): Water Scarcity and the Impact of the Mining and Agricultural Sectors in Chile. In *Sustainability* 8 (2), p. 128. DOI: 10.3390/su8020128.

Aldunce, Paulina; Bórquez, Roxana; Indvik, Katy; Lillo, Gloria (2015): Identificación de actores relacionados a la sequía en Chile. Julio, 2015. Edited by (CR)2 Centro de Ciencia del Clima y la Resiliencia.

Amnestía Internacional (2018): Chile: Las autoridades deben proteger a Rodrigo Mundaca y a otras personas que defienden el medio ambiente. Available online at https://www.amnesty.org/es/press-releases/2018/06/chile-authorities-must-protect-rodrigo-mundaca-and-other-environmental-defenders/, checked on 12/8/2018.

Andolina, Robert; Laurie, Nina; Radcliffe, Sarah A. (2009): Indigenous development in the Andes. Culture, power, and transnationalism. Durham: Duke University Press. Available online at http://site.ebrary.com/lib/ alltitles/docDetail.action?docID=10389685.

Australian Government (2018): ENSO Outlook. Edited by Commonwealth of Australia. Available online at http://www.bom.gov.au/climate/enso/outlook/ #tabs=ENSO-Outlook-history, updated on 2018, checked on 12/7/2018.

Baasch, Stefanie; Bauriedl, Sybille; Hafner, Simone; Weidlich, Sandra (2012): Klimaanpassung auf regionaler Ebene: Herausforderungen einer regionalen Klimawandel-Governance. In: *Raumforschung und Raumordnung* 70(3). 191 – 201.

Baechler, Günther (Ed.) (2002): Transformation of resource conflicts. Approach and instruments. Bern: Lang.

Bahadur, Aditya V.; Ibrahim, Maggie; Tanner, Thomas (2013): Characterising resilience. Unpacking the concept for tackling climate change and development. In *Climate and Development* 5 (1), 55–65. DOI: 10.1080/17565529.2012.762334.

Banco Mundial (2013): CHILE. Estudio para el mejoramiento del marco institucional para la gestión del agua. Available online at http://reformacodigodeaguas.carey.cl/ wp-content/uploads/2014/09/Informe-Banco-Mundial-Estudio-para-el-mejoramiento-del-marco-institucional.pdf, checked on 12/8/2018.

Banco Mundial (2011): CHILE: Diagnóstico de la gestión de los recursos hídricos. Departamento de Medio Ambiente y Desarrollo Sostenible Región para América Latina y el Caribe. Available online at: http://documentos.bancomundial.org/ curated/es/452181468216298391/Chile-Diagn-243-stico-de-la-gesti-243-n-de-los-recursos-h-237-dricos; checked on 14/06/2020

Barozet, Emmanuelle; Espinoza, Vicente (2016): Current Issues on the Political Representation of Middle Classes in Chile. In *Journal of Politics in Latin America* (3), 95–123.

Basurto, Xavier; Gelcich, Stefan; Ostrom, Elinor (2013): The social–ecological system framework as a knowledge classificatory system for benthic small-scale fisheries. In *Global Environmental Change* 23 (6), 1366–1380. DOI: 10.1016/j.gloenvcha.2013.08.001.

Bauer, Carl J. (2015): Canto de sirenas. El derecho de aguas chileno como modelo para reformas internacionales. With assistance of Juan Pablo Orrego Silva. 2nd. ed. Providencia, Santiago de Chile: El Desconcierto.

BBC Mundo (2016): Qué es la marea roja por la que declararon una catástrofe en Chile. Available online at https://www.bbc.com/mundo/noticias/2016/05/160503_chile_marea_roja_catastrofe_alerta_sanitaria_ab , checked on 11/5/2018.

BCN (2018a): La Ligua - Reportes Estadisticos Comunales. Edited by Biblioteca del Congreso Nacional de Chile. Available online at http://reportescomunales.bcn.cl/2015/index.php/La_Ligua, updated on 9/7/2018, checked on 12/7/2018.

BCN (2018b): Petorca - Reportes Estadisticos Comunales. Edited by Biblioteca del Congreso Nacional de Chile. Available online at http://reportescomunales.bcn.cl/2015/index.php/Petorca, updated on 9/7/2018, checked on 12/7/2018.

BCN (2018c): Cabildo - Reportes Estadisticos Comunales. Edited by Biblioteca del Congreso Nacional de Chile. Available online at http://reportescomunales.bcn.cl/2015/index.php/Cabildo, updated on 9/7/2018, checked on 12/7/2018.

Beckenkamp, Martin (2014): Der Umgang mit sozialen Dilemmata. Institutionen und Vertrauen in den Commons. In: Helfrich, Silke; Heinrich-Böll-Stiftung (Eds.): Commons. 1st ed. Bielefeld: transcript (Wirtschafts-, Organisations- und Arbeitssoziologie), 51–57.

Belmar, Antonio; Carvajal, Luis; Cortez, Manuel; Castro, Harris; Silva, Sandra; Ferreiro, Karla et al. (2010): Conflicts Over Water in Chile: Between Human Rights and Market Rules. Edited by Sara Larraín, Colombina Schaeffer. Chile Sustentable. Santiago de Chile.

Benedikter, Roland; Siepmann, Katja (2015): Chile in Transition. Cham: Springer International Publishing.

Berger, Peter L.; Luckmann Thomas (1993), Die gesellschaftliche Konstruktion der Wirklichkeit -Eine Theorie der Wissenssoziologie, 5th ed., Frankfurt a.M..

Binder, Claudia R.; Hinkel, Jochen; Bots, Pieter W. G.; Pahl-Wostl, Claudia (2013): Comparison of Frameworks for Analyzing Social-ecological Systems. In *E&S* 18 (4). DOI: 10.5751/ES-05551-180426.

Boelens, Rutgerd Anne (2008): THE RULES OF THE GAME AND THE GAME OF THE RULES. Normalization and resistance in Andean water control. Wageningen University. The Netherlands.

Boelens, Rutgerd; Cremers, Leontien; Zwarteveen, Margreet (2011): Justicia Hídrica: acumulación de agua, conflictos y acción de la sociedad civil. In: Boelens, Rutgerd; Cremers, Leontien; Zwarteveen, Margreet (Eds): Justicia hídrica: acumulación, conflicto y acción social. Lima: IEP; Fondo Editorial PUCP, Justicia Hídrica. 2011. (Agua y Sociedad, 15. Serie Justicia Hídrica, 1).

Boisier, Juan P.; Rondanelli, Roberto; Garreaud, René D.; Muñoz, Francisca (2016): Anthropogenic and natural contributions to the Southeast Pacific precipitation decline and recent megadrought in central Chile. In *Geophys. Res. Lett.* 43 (1), 413–421. DOI: 10.1002/2015GL067265.

Bolados, García Paola (2016): Conflictos socio-ambientales/territoriales y el surgimiento de identidades post neoliberales (Valparaíso-Chile). Social and environmental / territorial conflicts and the emergence of post neoliberals identities (Valparaíso-Chile). In *Izquierdas* (31), 102–129.

Bonnefoy, Pascale (2016): Daughter-in-Law of Chile's President Faces Corruption Charge. Edited by The New York Times. Available online at https://www.nytimes.com/2016/01/30/world/americas/daughter-in-law-of-chiles-president-charged-with-tax-crimes-and-bribery.html, checked on 11/5/2018.

Briscoe, John (1996): Water as an Economic Good: The Idea and what it means in Practice, Cairo (World Congress of the International Commission on Irrigation and Drainage).

Brunnengräber, Achim (2011): Zivilisierung des Klimaregimes. NGOs und soziale Bewegungen in der nationalen, europäischen und internationalen Klimapolitik. 1st ed. Wiesbaden: VS Verlag für Sozialwissenschaften / Springer Fachmedien Wiesbaden GmbH Wiesbaden (Energiepolitik und Klimaschutz). Available online at http://dx.doi.org/10.1007/978-3-531-92840-1.

Budds, Jessica (2012): La demanda, evaluación y asignación del agua en el contexto de escasez. Un análisis del ciclo hidrosocial del valle del río La Ligua, Chile. In *Rev. geogr. Norte Gd.* (52), 167–184. DOI: 10.4067/S0718-34022012000200010.

Bustamante Pizarro, Ricardo (2018): Rodrigo Mundaca: »Los pilares de la desigualdad en Chile están edificados sobre la base de la apropiación masiva de los bienes comunes« | Causas y Beats. Available online at https://www.causasybeats.cl/ movimiento-social/rodrigo-mundaca-los-pilares-de-la-desigualdad-en-chile-estan-edificados-sobre-la-base-de-la-apropiacion-masiva-de-los-bienes-comunes/, checked on 12/8/2018.

Carmona-Lopez, Alejandra (2018): Palteros de Petorca al banquillo: las críticas al modelo chileno que deja pobreza y desigualdad. El Mostrador (Ed). Available online at https://www.elmostrador.cl/noticias/pais/2018/05/25/palteros-de-petorca-al-banquillo-las-criticas-al-modelo-chileno-que-deja-pobreza-y-desigualdad/, checked on 12/7/2018.

Castro L., Milka (2007): 'Comunidad territorial indígena, gestión de recursos hídricos y derechos colectivos'. In: Boelens, Roy; Guevara, Armando; Hendriks, Jasper. and Hoogesteger (eds.), Pluralismo legal, reforma hídrica y políticas de reconocimiento, 143-156. WALIR Studies vol. 13. Wageningen / Cusco: UN/CEPAL and Wageningen University.

CIA (Central Intelligence Agency) (2013): The World Factbook — Central Intelligence Agency. Available online at https://www.cia.gov/library/publications /the-world-factbook/rankorder/2172rank.html, updated on 11/14/2018, checked on 12/9/2018.

Clarvis, Margot Hill; Allan, Andrew (2014): Adaptive capacity in a Chilean context: A questionable model for Latin America. In *Environmental Science & Policy* 2014 (43), 78–90. Available online at http://dx.doi.org/10.1016/j.envsci.2013.10.014, checked on 6/24/2015.

Cleaver, Frances (2012) Development through Bricolage: Rethinking Institutions for Natural Resource Management, London: Routledge.

Cleaver, Frances; de Koning, Jessica (2015): Furthering Critical Institutionalism. In *International Journal of the Commons* (9), 1–18. Available online at http://www.thecommonsjournal.org.

Cleaver, Frances; Whaley, Luke (2018): Understanding process, power, and meaning in adaptive governance. A critical institutional reading. In *E&S* 23 (2). DOI: 10.5751/ES-10212-230249.

CNR and Universidad de Concepción (2016): Estudio básico diagnóstico para desarrollar plan de riego en las cuencas de los ríos La Ligua y Petorca. Informe final. TOMO I-III. Edited by CNR Ministerio de Agricultura. Departamento de Recursos Hídricos Facultad de Ingeniería Agrícola. Available online at: http://biblioteca digital.ciren.cl/handle/123456789/26768, checked on 22/06/2020.

Corporación Latinobarómetro (2018): Latinobarómetro Database. Santiago de Chile. Available online at http://www.latinobarometro.org/latOnline.jsp, checked on 12/7/2018.

Dawson, Patrick; Daniel, Lisa (2010): Understanding social innovation. A provisional framework. In *IJTM* 51 (1), 9. DOI: 10.1504/IJTM.2010.033125.

del Mar Delgado-Serrano, María; Ramos, Pablo Andrés (2015): Making Ostrom's framework applicable to characterise social ecological systems at the local level. In *International Journal of the Commons* 2015 (2), 808–830. Available online at http://www.thecommonsjournal.org.

della Porta, Donnatella (2008): Comparative analysis: case-oriented versus variable-oriented research. In: Della Porta, Donatella; Keating, Michael (Eds.): Approaches and Methodologies in the Social Sciences. A Pluralist Perspective. Leiden: Cambridge University Press, 198–222.

Deutsch, Morton (1975): Equity, Equality, and Need. What Determines Which Value Will Be Used as the Basis of Distributive Justice? In *Journal of Social Issues* 31 (3), 137–149. DOI: 10.1111/j.1540-4560.1975.tb01000.x.

Deutsch, Morton (1985): Distributive justice. A social-psycholog. perspective. New Haven: Yale Univ. Pr.

DGA (2018): Áreas de Restricción de Aguas Subterráneas. Edited by Ministerio de Obras Públicas - Dirección de General de Aguas. Available online at http://www.dga.cl/administracionrecursoshidricos/areasderestriccion/Paginas/defaul t.aspx, updated on 12/8/2018, checked on 12/8/2018.

References

Diario El Divisadero (2017): Director DGA: "Quien saca aguas sin el derecho a sacarla tiene sanciones mínimas". Available online at http://www.eldivisadero.cl/redac-42880, checked on 12/8/2018.

Donoso, Guillermo; Vicuña, Sebastian (2016): Aseguramiento responsable del agua para consumo humano. Technical Report.

Durán, Gonzalo; Kremerman, Marco (2016): Los Verdaderos Sueldos de la Región de Valparaíso. Panorama Actual del Valor del Trabajo usando la Encuesta Suplementaria de Ingresos 2015. Fundación Sol.

Eid, Ursula; Kranz, Nicole (2014): Wasser: Menschenrecht, Ressource, Konfliktstoff? In: Schneckener, Ulrich; von Scheliha, Arnulf; Lienkamp, Andreas; Klagge, Britta (Eds.): Wettstreit um Ressourcen. Konflikte um Klima, Wasser und Boden. Oekom. München. 139 – 156

EL CIUDADANO (2014): Dirigente es condenado por defender el agua en Chile. La justicia chilena condenó a Rodrigo Mundaca de Modatima, provincia de Petorca, a 541 días de cárcel por denunciar usurpación del agua contra Edmundo Perez-Yoma, ex-ministro de Michelle Bachelet. elciudadano.com. Available online at https://www.elciudadano.cl/medio-ambiente/dirigente-es-condenado-por-defender-el-agua-en-chile/04/08/, checked on 12/8/2018.

el periodista (2018): Diputados denuncian intento por privatizar Dirección General de Aguas (DGA) El Periodista Online.). Available online at https://www. elperiodista.cl/economia/2018/11/diputados-denuncian-intento-por-privatizar-direccion-general-de-aguas-dga/, checked on 12/8/2018.

El siglo (2018): Se frenó el intento de privatizar la DGA. Available online at http://www.elsiglo.cl/2018/11/29/se-freno-el-intento-de-privatizar-la-dga/, checked on 12/8/2018.

Espinoza, Vicente (2013): Local associations in Chile: social innovations in a mature neoliberal society. In: Moulaert, Frank (Ed.): The international handbook on social innovation. Collective action, social learning and transdisciplinary research. Cheltenham: Elgar, 387–411.

Evans, J. P. (2011): Resilience, ecology and adaptation in the experimental city. In *Transactions of the Institute of British Geographers* 36 (2), 223–237. DOI: 10.1111/j.1475-5661.2010.00420.x.

Fiksel, Joseph (2006): Sustainability and resilience: toward a systems approach. Center for Resilience. Columbus (USA), 14-21.

Folke, Carl; Carpenter, Steve; Elmqvist, Thomas; Gunderson, Lance; Holling, C. S.; Walker, Brian (2002): Resilience and sustainable development. Building adaptive capacity in a world of transformations. In *Ambio* 31 (5), 437–440.

Franklin, Jonathan (2015): Chilean president rocked by corruption allegations against family members. Michelle Bachelet denies resignation rumors as her son Sebastian Davalos is to be questioned about his role in a possibly illegal real estate transaction. Edited by The Guardian. Santiago de Chile. Available online at https://www.theguardian.com/world/2015/apr/08/chilean-president-michelle-bachelet-corruption-charges-sebastian-davalos, checked on 11/5/2018.

Franziskus (2015): Die Enzyklika »Laudato si'« über die Sorge für das gemeinsame Haus. [die Umwelt-Enzyklika des Papstes]. Freiburg im Breisgau: Herder.

Fuster, Rodrigo; Donoso, Guillermo (2018): Rural Water Management. In: Donoso, Guillermo (Ed.): Water Policy in Chile. Cham: Springer International Publishing (Global Issues in Water Policy, 21), 151–164.

Garreaud, René D.; Alvarez-Garreton, Camila; Barichivich, Jonathan; Boisier, Juan Pablo; Duncan, Christie; Galleguillos, Mauricio; LeQuesne, Carlos; McPhee, James; Zambrano-.Bigiarini, Mauricio (2017): The 2010–2015 megadrought in central Chile. Impacts on regional hydroclimate and vegetation. In *Hydrol. Earth Syst. Sci.* 21 (12), 6307–6327. DOI: 10.5194/hess-21-6307-2017.

Getches, David H. (2005) 'Defending Indigenous Water Rights with the Laws of a Dominant Culture: the Case of the United States.' In: Liquid Relations. Contested Water Rights and Legal Complexity, Roth, Dik; Boelens, Rutgerd; Zwarteveen, Margreet (Eds.), 44-65. New Brunswick, New Jersey, London: Rutgers University Press.

Gobernación, Provincia de Petorca (2016): Resumen de Estrategia Provincial de Petorca. Pontificia Universidad Católica de Valparaíso. CEAL. Centro de Estudios y Asistencia Legislativa. Ministerio del Interior y Seguridad Pública.

Gobierno de Chile (2010): Región de Valparaíso 2020. Obras públicas para el desarrollo. Available online at http://www.dirplan.cl/planes/regionales/Documents/V/MOP%20SEPARATA%20VALPARAISO%2018-12.pdf, checked on 12/8/2018.

Gobierno de Chile (2018): Gobierno de Chile - Directorio de Transparencia Activa - Acerca De. Available online at http://www.gobiernotransparentechile.gob.cl/ pagina/acercade, checked on 12/7/2018.

González, Sara; Moulaert, Frank; Martinelli, Flavia (2010): ALMOLIN. How to analyse social innovation at the local level? In: Moulaert, Frank (Ed.): Can neighbourhoods save the city? Community development and social innovation. 1. publ. London u.a.: Routledge (Regions and cities).

Gopalakrishnan, Chennat (2005): Water Allocation and Management in Hawaii: A Case of Institutional Entropy. In: Gopalakrishnan, Chennat; Biswas, Asit K.; Tortajada, Cecilia (Eds.): Water Institutions: Policies, Performance and Prospects. Berlin, Heidelberg: Springer-Verlag Berlin Heidelberg (Water Resources Development and Management), 1–23.

Hall, Kurt; Cleaver, Frances; Franks, Tom; Maganga, Faustin (2013): Critical institutionalism: a synthesis and exploration of key themes. In: *Environment, Politics and Development Working Paper Series* (63), 1–41. Available online at http://www.kcl.ac.uk/sspp/departments/geography/research/epd/wp63Cleaver.pdf.

Hall, Peter A.; Taylor, Rosemary C. R. (1996): Political Science and the Three New Institutionalisms. In *MPIFG Discussion Paper* 96 (6), 1–32.

Hardin, Garrett (1968): The tragedy of the commons. In *Science (New York, N.Y.)* 162 (3859), 1243–1248. DOI: 10.1126/science.162.3859.1243.

Hayek, Friedrich A. von (1999): The road to serfdom. 50. anniversary ed., [Nachdr.]. Chicago: Univ. of Chicago Press.

Heinrichs, Harald; Grunenberg, Heiko (2009): Klimawandel und Gesellschaft. Perspektive Adaptionskommunikation. VS Verlag für Sozialwissenschaften, Wiesbaden.

Henríquez, Cristián; Aspee, Nicolle; Quense, Jorge (2016): Zonas de catástrofe por eventos hidrometeorológicos en Chile y aportes para un índice de riesgo climático. In *Rev. geogr. Norte Gd.* (63), 27–44. DOI: 10.4067/S0718-34022016000100003.

Hoering, Uwe (2006): Der Markt als Wassermanager. Aufbau eines neuen Wasserregimes durch die Weltbank. In: Peripherie 101/102, 21-42.

Höld, Regina (2009): Zur Transkription von Audiodaten. In: Buber, Renate; Holzmüller, Hartmut H. (Eds.): Qualitative Marktforschung. Konzepte - Methoden - Analysen. 2., überarbeitete Auflage. Wiesbaden: Gabler Verlag / GWV Fachverlage GmbH Wiesbaden (Lehrbuch), 655–668.

Holling, C. S.; Gunderson, Lance H. (2002): Resilience and Adaptive Cycles. In: Lance H. Gunderson, Crawford S. Holling (Eds.): Panarchy. Understanding transformations in human and natural systems. Washington, DC: Island Press.

Houdret, Annabelle (2008): Knappes Wasser, reichlich Konflikte? Lokale Wasserkonflikte und die Rolle der Entwicklungszusammenarbeit. Institut für Entwicklung und Frieden. Duisburg.

Houdret, Annabelle (2010): Wasserkonflikte sind Machtkonflikte. Ursachen und Lösungsansätze in Marokko. Paris, Univ., Diss.--Duisburg-Essen, 2008. Wiesbaden: VS Verl. für Sozialwiss. Available online at http://dx.doi.org/10.1007/978-3-531-92318-5.

Houdret, Annabelle; Shabafrouz, Miriam (2006): Privatisation in Deep Water? . Water Governance and Options for Development Cooperation. Edited by Institute for Development and Peace. University of Duisburg-Essen. Duisburg (84).

Hyde, Kylie (2014): Water Footprinting in the Urban Water Sector. London: IWA Publishing (Gwrc report series).

INDAP (2017): Agua para Petorca: Inauguran tranque en Cabildo y entregan derechos de agua en Longotoma. Edited by INDAP. Available online at http://www.indap .gob.cl/noticias/detalle/2017/12/04/agua-para-petorca-inauguran-tranque-en-cabildo-y-entregan-derechos-de-agua-en-longotoma, updated on 2017, checked on 12/7/2018.

INDH (2012): Instituto Nacional de Derechos Humanos: Mapa de conflictos socioambientales en Chile. Usurpación de aguas en Petorca, Cabildo y La Ligua. Available online at: https://bibliotecadigital.indh.cl/bitstream/handle/123456789/478/mapa-conflictos.pdf?sequence=4, checked on 07/12/2018, 142 – 144.

IPCC (2007): Climate Change 2007: Synthesis Report. Contribution of Working Groups I, II and III to the Fourth Assessment Report of the Intergovernmental Panel on Climate Change [Core Writing Team, Pachauri, R.K and Reisinger, A. (eds.)]. IPCC, Geneva, Switzerland, 104 pp.

IPCC (2014): Climate Change 2014: Synthesis Report. Contribution of Working Groups I, II and III to the Fifth Assessment Report of the Intergovernmental Panel on Climate Change [Core Writing Team, R.K. Pachauri and L.A. Meyer (eds.)]. IPCC, Geneva, Switzerland, 151 pp.

IPCC (2015): Climate Change 2014: Impacts, Adaptation and Vulnerability, Volume 2, Regional Aspects. Working Group II Contribution to the IPCC Fifth Assessment Report. Cambridge: Cambridge University Press. Available online at http://dx.doi.org/10.1017/CBO9781107415386.

IPCC (2018): Global Warming of 1,5°C. An IPCC Special Report on the impacts of global warming of 1.5°C above pre-industrial levels and related global greenhouse gas emission pathways, in the context of strengthening the global response to the threat of climate change, sustainable development, and efforts to eradicate poverty. Edited by Inergovernmental Panel on Climate Change.

Johnson, Craig (2004): 'Uncommon Ground: The 'Poverty of History' in Common Property Discourse', Development and Change, 35(3): 407-33.

Jones, Stephen (2013): Sharing the recurrent costs of rural water supply in Mali: the role of WaterAid in promoting sustainable service delivery: Oxford University Press.

Kanol, Direnç (2015): Tutelary Democracy in Unrecognized States (6). In: *EUL Journal of Social Sciences*. LAÜ Sosyal Bilimler Dergisi June 2015 Haziran. Available online at: http://euljss.eul.edu.tr/euljss/si614.pdf, checked on 22/06/2020

Kiser, Larry; Ostrom, Elinor (1982): Strategies of Political Inquiry. Beverly Hills, CA: Sage.

Komakech, Hans, Van Koppen, Barbara, Mahoo, Henry & Van Der Zaag, Pieter (2011) 'Pangani River Basin over Time and Space: On the Interface of Local and Basin Level Responses', Agricultural Water Management, 98(11): 1740-1751.

Larraín, Sara (2010): Introduction: Between Human Rights and Market Rules. In Sara Larraín, Colombina Schaeffer (Eds.): Conflicts Over Water in Chile: Between Human Rights and Market Rules. Santiago de Chile, 5–32.

Lindenberg, Siegwart (1991): Die Methode der abnehmenden Abstraktion: Theoriegesteuerte Analyse und empirischer Gehalt. In: Esser, Hartmut; Troitzsch; Klaus (Eds.), Modellierung sozialer Prozesse. Bonn, 29–78.

Lobina, Emanuele; Kishimoto, Satoko; Petitjean, Oliver (2014): Here to stay: water remunicipalisation as a global trend. 180 cases in 2014. Edited by Public Services International Research Unit (PSIRU), Transnational Institute (TNI), Multinational Observatory.

Maddocks, Andrew; Young, Robert Samuel; Reig, Paul (2015): Ranking the World's Most Water-Stressed Countries in 2040 | World Resources Institute. Edited by World Resources Institute. Available online at https://www.wri.org/blog/2015/08/ranking-world-s-most-water-stressed-countries-2040, checked on 12/7/2018.

Mahoney, James (2012): The Logic of Process Tracing Tests in the Social Sciences. In *Sociological Methods & Research* 41 (4), 570–597. DOI: 10.1177/0049124112437709.

March, James G.; Olsen, Johan P. (1984) »The New Institutionalism: Organizational Factors in Political Life.« The American Political Science Review. Vol. 78. No.3. 734-749.

Mayntz, Renate; Scharpf, Fritz W. (1995): Der Ansatz des akteurzentrierten Instutionalismus. In: Mayntz, Renate; Scharpf, Fritz W. (Eds.): Gesellschaftliche Selbstregelung und politische Steuerung. Frankfurt am Main: Campus-Verl. (Schriften des Max-Planck-Instituts für Gesellschaftsforschung Köln, 23), 39–72.

Mayring, Philipp (2010): Qualitative Inhaltsanalyse. Grundlagen und Techniken. 11., aktualisierte und überarb. Aufl. Weinheim u.a.: Beltz (Pädagogik).

McGinnis, Michael D.; Ostrom, Elinor (2014): Social-ecological system framework. Initial changes and continuing challenges. In: *E&S* 19 (2). DOI: 10.5751/ES-06387-190230.

Merkel, Wolfgang (2004): Embedded and defective democracies. In *Democratization* 11 (5), 33–58. DOI: 10.1080/13510340412331304598.

Merkel, Wolfgang (2010): Systemtransformation. Eine Einführung in die Theorie und Empirie der Transformationsforschung. 2., überarb. und erw. Aufl. Wiesbaden: VS Verl. für Sozialwiss (Lehrbuch).

Messner, Dirk (1997): The network society. Economic development and international competitiveness as problems of social governance. London: Cass (GDI book series, 10). Available online at http://www.loc.gov/catdir/enhancements/fy0652/97023233-d.html.

Messner, Dirk (2012): Elinor Ostrom und James Walker, Trust and Reciprocity. In: Leggewie, Claus; Zifonun, Dariuš; Lang, Anne; Siepmann, Marcel; Hoppen; Johanna (Eds.): Schlüsselwerke der Kulturwissenschaften. Bielefeld: transcript (Edition Kulturwissenschaft, Band 7), 241–244.

Messner, Dirk; Guarín, Alejandro; Haun, Daniel (2016): The behavioral dimensions of international cooperation. In: Messner, Dirk; Weinlich, Silke (Eds.): Global cooperation and the human factor in international relations. London, New York: Routledge (Routledge global cooperation series), 47–65.

Messner, Dirk; Guarín, Alejandro; Haun, Daniel (2013): The Behavioural Dimensions of International Cooperation. In: SSRN Journal. DOI: 10.2139/ssrn.2361423.

Meuleman, Louis (2008): Public Management and the Metagovernance of Hierarchies, Networks and Markets. The Feasibility of Designing and Managing Governance Style Combinations. Zugl.: Rotterdam, Univ., Diss., 2008. Heidelberg, Berlin, Heidelberg: Physica-Verl.; Springer (Contributions to Management Science). Available online at http://d-nb.info/989410692/34.

Ministerio de Obras Públicas (2018a): Ministerio de obras públicas - Coordinación de Concesiones de Obras Públicas. Available online at http://www.concesiones.cl/proyectos/Paginas/detalle_adjudicacion.aspx?item=192, checked on 12/7/2018.

Ministerio de Obras Públicas (2018b): Ministerio de obras públicas - Coordinación de Concesiones de Obras Públicas. Available online at http://www.concesiones.cl/proyectos/Paginas/detalle_adjudicacion.aspx?item=196, checked on 12/7/2018.

Molinos-Senante, María (2018): Urban Water Management. In: Donoso, Guillermo (Ed.): Water Policy in Chile. Cham: Springer International Publishing (Global Issues in Water Policy, 21), 131–150.

Montes, Rocío (2015): Chile breaks up toilet paper cartel that fixed prices for over a decade. But lax legislation means that the country's top two manufacturers will likely get off lightly. Edited by El País. Santiago de Chile. Available online at https://elpais.com/elpais/2015/10/30/inenglish/1446199130_705491.html, checked on 11/5/2018.

MOP (2017): Registro Público de Organizaciones de Usuarios. Edited by Ministerio de Obras Públicas - Dirección de General de Aguas (.). Available online at http://www.dga.cl/administracionrecursoshidricos/OU/Paginas/default.aspx, checked on 12/8/2018.

MOP; DGA (2015): Atlas del Agua. Chile 2016. Edited by Ministerio de Obras Públicas. Santiago de Chile.

Moulaert, Frank (Ed.) (2013): The international handbook on social innovation. Collective action, social learning and transdisciplinary research. Cheltenham: Elgar.

Murray, Robin; Caulier-Grice, Julie; Mulgan, Geoff (2010): The open book of social innovation (Social Innovator Series: Ways To Design, Develop And Grow Social Innovation). The Young Foundation. Nesta. Innovating public services.

Mwangi, Esther; Markelova, Helen (2014): Lokal, regional, global? Mehrebenen-Governance und die Frage des Maßstabs. In Helfrich, Silke; Heinrich-Böll-Stiftung (Eds.): Commons. 1st ed. Bielefeld: transcript (Wirtschafts-, Organisations- und Arbeitssoziologie), 455–465.

Naciones Unidas, Gobierno de Chile (2010): Objetivos de Desarrollo del Milenio. Tercer Informe del Gobierno de Chile. Available online at: http://www.undp.org /content/dam/undp/library/MDG/english/MDG%20Country%20Reports/Chile/Terc er%20Informe%20Nacional%20ODM%20Chile%202010.pdf, checked on 20/04/2014) 118 – 147.

North, Douglass Cecil (1990): Institutions, Institutional Change, and Economic Performance. Cambridge University Press. 3.

North, Douglass Cecil (1993): Institutions and Credible Commitment, in: Journal of Institutional and Theoretical Economics (JITE), 149.1, 11-23.

ODEPA (2018): Empleo agrícola - ODEPA | Oficina de Estudios y Políticas Agrarias. Available online at https://www.odepa.gob.cl/estadisticas-del-sector/empleo-agricola, checked on 12/7/2018.

ODEPA Oficina de Estudios y Políticas Agrarias - Office of Agricultural Studies and Policies (2017): Panorama de la Agricultura Chilena. Chilean Agriculture Overview 2017. Available online at https://www.odepa.gob.cl/wp-content/uploads/2017/12/ panoramaFinal20102017Web.pdf, checked on 12/7/2018.

OECD (2005): Environmental Performance Reviews. CHILE. OECD. Paris.

OECD (2011): Society at a Glance 2011. OECD Social Indicators. 1st ed. s.l.: OECD. Available online at http://gbv.eblib.com/patron/FullRecord.aspx?p=714228.

OECD (2015): Water Resources Allocation: Sharing Risks and Opportunities, OECD Studies on Water, OECD Publishing, Paris, https://doi.org/10.1787/9789264229631-en.

OECD/ECLAC (2016): OECD Environmental Performance Reviews: Chile 2016, OECD Environmental Performance Reviews, OECD Publishing, Paris, https://doi.org/10.1787/9789264252615-en.

OECD Observer (2010): Climate change and agriculture. Available online at http://oecdobserver.org/news/archivestory.php/aid/3213/Climate_change_and_agriculture.html, checked on 12/8/2018.

Ostrom, Elinor (1990): Governing the commons. The evolution of institutions for collective action. 1. publ. Cambridge Mass. u.a.: Cambridge Univ. Press (Political economy of institutions and decisions).

Ostrom, Elinor (2005): Toward a Behavioral Theory Linking Trust, Reciprocity, and Reputation. In: Ostrom, Elinor; Walker, James (Eds.): Trust and reciprocity. Interdisciplinary lessons from experimental research. 1. papercover ed. New York: Russell Sage Foundation (Russell Sage Foundation series on trust, 6), 19–79.

Ostrom, Elinor (2005a): Understanding institutional diversity. Princeton, New Jersey, Oxford: Princeton University Press (Princeton paperbacks). Available online at http://www.esmt.eblib.com/patron/FullRecord.aspx?p=483578.

Ostrom, Elinor (2007): A diagnostic approach for going beyond panaceas. In: *Proceedings of the National Academy of Sciences of the United States of America* 104 (39), 15181–15187. DOI: 10.1073/pnas.0702288104.

Ostrom, Elinor (2008): Design Principles Of Robust Property-Rights Institutions: What Have We Learned? . Workshop in Political Theory and Policy Analysis. Indiana University. Cambridge, 7/21/2008. Available online at http://poseidon01. ssrn.com/delivery.php?ID=49409612000502717100410612512709609202401708608602904902511901808512711909312708502800312312202605100602308809407311511307807604304707403000302702810709611100009400100504907402906412710208902509809512706910607511911408110911811911500203011108700 1&EXT=pdf&TYPE=2, checked on 5/15/2015.

Ostrom, Elinor (2009): A general framework for analyzing sustainability of social-ecological systems. In: *Science (New York, N.Y.)* 325 (5939), 419–422. DOI: 10.1126/science.1172133.

Ostrom, Elinor (Ed.) (2003): Trust and reciprocity. Interdisciplinary lessons from experimental research. New York NY: Russell Sage Foundation (The Russell Sage Foundation series on trust, 6).

Ostrom, Elinor; Cox, Michael (2010): Moving beyond panaceas. A multi-tiered diagnostic approach for social-ecological analysis. In: *Envir. Conserv.* 37 (04), 451–463. DOI: 10.1017/S0376892910000834.

Ostrom, Elinor; Schöller, Ekkehard (1999): Die Verfassung der Allmende. Jenseits von Staat und Markt. Tübingen: Mohr Siebeck (Die Einheit der Gesellschaftswissenschaften, 104).

Ostrom, Elinor; Walker, James (Eds.) (2005): Trust and reciprocity. Interdisciplinary lessons from experimental research. Roundtable on Trust as a Political Variable; Conference on Trust to meet at the Russell Sage Foundation; Conference on Experimental Work Relating to the Study of Trust. 1. papercover ed. New York: Russell Sage Foundation (Russell Sage Foundation series on trust, 6).

Paavola, Jouni; Adger, Neil W.; Huq, Saleemul (2006): Multifaced Justice in Adaptation to Climate Change. In: Adger, Neil; Paavola, Jouni; Huq, Saleemul; Mace, M.J. (2006): Fairness in Adaptation to Climate Change. Cambrige/ London: MIT Press.

Pahl-Wostl, C. (2009). A conceptual framework for analysing adaptive capacity and multi-level learning processes in resource governance regimes. Global Environmental Change 19:354-365. http://dx.doi.org/10.1016/j.gloenvcha.2009.06.001.

Pahl-Wostl, Claudia; Kranz, Nicole (2010). Editorial to special issue: Water governance in times of change. Environmental Science & Policy 13:567-570. http://dx.doi.org/10.1016/j.envsci.2010.09.004.

Pahl-Wostl, Claudia (2015): Water Governance in the Face of Global Change. From Understanding to Transformation. 1st ed. 2015. Cham: Springer (Water Governance - Concepts, Methods, and Practice). Available online at http://search.ebscohost.com/login.aspx?direct=true&scope=site&db=nlebk&AN=1060624.

Pahl-Wostl, Claudia; Holtz, Georg; Kastens, Britta; Knieper, Christian (2010): Analyzing complex water governance regimes: the Management and Transition Framework. In: *Environmental Science & Policy* 2010 (13), 571–581.

Peña, Humberto (2018): Integrated Water Resources Management in Chile: Advances and Challenges. In: Donoso, Guillermo (Ed.): Water Policy in Chile. Cham: Springer International Publishing (Global Issues in Water Policy, 21), 197–207.

Perry, C.J.; Rock, Michael; Seckler, David (1997): Water as an Economic Good: A Solution, or a Problem? Colombo (International Irrigation Management Institute IIMI, Research Report 14).

Peters, B. Guy (2005) Institutional Theory in Political Science: The 'New Institutionalism'. Continuum International Publishing Group.

Pew Research Center (2017): Public Trust in Government: 1958-2017. Available online at http://www.people-press.org/2017/12/14/public-trust-in-government-1958-2017/, updated on 2017, checked on 12/7/2018.

Pfeiffer, Evelyn (2016): Chile's Record Toxic Tides May Have Roots in Dirty Fish Farming. A red tide crisis has stirred fishermen to unrest and may be at least partially linked to controversial aquaculture practices, while El Niño is also to blame. Edited by National Geographic. Available online at https://news.nationalgeographic.com/2016/05/160517-chile-red-tide-fishermen-protest-chiloe/, checked on 11/5/2018.

Pigg, Stacy Leigh (1996) 'The Credible and the Credulous: The Question Of »Villagers' Beliefs« In Nepal', Cultural Anthropology, 11(2): 160-201.

Pol, Eduardo; Ville, Simon (2009): Social innovation. Buzz word or enduring term? In: The Journal of Socio-Economics 38 (6), 878–885. DOI: 10.1016/j.socec.2009.02.011.

Poteete, Amy R.; Janssen, Marco A.; Ostrom, Elinor (2010): Working together. Collective action, the commons, and multiple methods in practice. Princeton, NJ: Princeton Univ. Press.

Pradel Miquel, Marc; García Cabeza, Marisol; Eizaguirre Anglada, Santiago (2013): Theorizing multi-level governance in social innovation dynamics. In: Frank Moulaert (Ed.): The international handbook on social innovation. Collective action, social learning and transdisciplinary research. Cheltenham: Elgar, 155–168.

PUCV (2018): Programa - Gestión Hídrica Petorca. Edited by Pontificia Universidad Católica de Valparaíso. Available online at https://www.gestion hidricapetorca.cl/programa?p=legal, checked on 12/8/2018.

Retamal, Rafaela; Andreoli, Andrea; Arumi, José L.; Rojas, Jorge; Parra, Oscar (2013): Gobernanza Del Agua Y Cambio Climático: Fortalezas Y Debilidades Del Actual Sistema De Gestión Del Agua En Chile. Análisis Interno. In *Interciencia* 2013 (38), 8–16. Available online at http://www.interciencia.org/v38_01/008.pdf, checked on 6/24/2015.

Ríos, Marcela (2017): Diagnóstico sobre la Participación Electoral en Chile. Proyecto Fomentando la Participación Electoral en Chile. Edited by Programa De Las Naciones Unidas Para El Desarrollo (PNUD). Santiago de Chile.

Rivera, Andrés; Acuña, Cesar; Casassa, Gino; Bown, Francisca (2002): Use of remotely sensed and field data to estimate the contribution of Chilean glaciers to eustatic sea-level rise. In *Ann. Glaciol.* 34, 367–372. DOI: 10.3189/172756402781817734.

Rivera, Diego; Godoy-Faúndez, Alex; Lillo, Mario; Alvez, Amaya; Delgado, Verónica; Gonzalo-Martín, Consuelo et al. (2016): Legal disputes as a proxy for regional conflicts over water rights in Chile. In: *Journal of Hydrology* 535, 36–45. DOI: 10.1016/j.jhydrol.2016.01.057.

Romero, Simon (2015): Chile Joins Other Latin American Nations Shaken by Scandal. Edited by The New York Times. Available online at https://www.nytimes.com /2015/04/10/world/americas/chile-joins-other-latin-american-nations-shaken-by-scandal.html?_r=0&module=inline, checked on 11/5/2018.

Romzek, Janice C. (2009): The Machinery of Iraqi Water Institutions: The Development of Institutions in a Post War State. Lund University. Department of Political Science. Development Studies.

Roth, Dik; Boelens, Rutgerd; Zwarteveen, Margreet (eds.) (2005): Liquid Relations. Contested Water Rights and Legal Complexity. New Brunswick, New Jersey, London: Rutgers University Press.

Saleth, Maria R.; Dinar, Ariel (2005): Water institutional reforms: theory and practice. In Water Policy 2005 (7), 1–19. Available online at http://www.environmental- expert.com/Files%5C5302%5Carticles%5C9967%5CWaterinstitutionalreforms.pdf, checked on 5/17/2015.

Saretzki, Thomas (2010): Umwelt- und Technikkonflikte: Theorien, Fragestellungen, Forschungsperspektiven. In: Fendth, Perter H.; Saretzki, Thomas (Eds.): Umwelt- und Technikkonflikte. VS Verlag für Sozialwissenschaften. Wiesbaden. 33-35.

Scharpf, Fritz W. (1997): Games real actors play. Actor-centered institutionalism in policy research. Boulder, Colo.: Westview Press (Theoretical lenses on public policy). Available online at http://www.loc.gov/catdir/enhancements/fy0832/97016122-b.html.

Scharpf, Fritz W. (2006): Interaktionsformen. Akteurzentrierter Institutionalismus in der Politikforschung. Unveränd. Nachdr. der 1. Aufl. Wiesbaden: VS Verl. für Sozialwiss.

Schneidewind, Uwe; Feindt, Peter Henning; Meister, Hans Peter; Minsch, Jürg; Schulz, Tobias; Tscheulin, Jochen (1997): Institutionelle Reformen für eine Politik der Nachhaltigkeit: Vom Was zum Wie in der Nachhaltigkeitsdebatte. in: GAIA 6/3 (1997). 182-196.

Schulze, Holger (1997): Neo-Institutionalismus. Ein analytisches Instrument zur Erklärung gesellschaftlicher Transformationsprozesse. In: Arbeitspapiere des Bereichs Politik und Gesellschaft. Heft 4/1997. Osteuropa-Institut der Freien Universität Berlin. Arbeitsbereich Politik und Gesellschaft (Ed.): Klaus Segbers. Berlin.

TEPSIE (2014): Social innovation theory and research: a guide for researchers. A deliverable of the project: »The theoretical, empirical and policy foundations for building social innovation in Europe« (TEPSIE), European Commission – 7th Framework Programme, Brussels: European Commission, DG Research.

The Guardian (Ed.) (2015): Chile flushes out decade-long conspiracy to fix the price of toilet paper. Two companies formed an 'outrageous' cartel to control prices of a range of paper products which 'hurt poor people the most', says minister. Available online at https://www.theguardian.com/world/2015/oct/30/chile-flushes-out-decade-long-conspiracy-to-fix-the-price-of-toilet-paper, checked on 11/5/2015.

The Guardian (Ed.) (2016): Unprecedented 'red tide' crisis deepens in Chile's fishing-rich waters. The 'red tide' algal bloom, which turns the sea water red and makes seafood toxic, is believed to be one of the country's worst recent environmental crises. Available online at https://www.theguardian.com/world/2016/may/11/red-tide-crisis-deepens-in-chile-fishing-waters, checked on 11/5/2018.

The World Bank (2015): GINI index (World Bank estimate). Available online: http://data.worldbank.org/indicator/SI.POV.GINI, checked on 02/01/2015).

The World Bank Group (2017): Worldwide Governance Indicators. Available online at http://info.worldbank.org/governance/wgi/index.aspx#reports, checked on 4/29/2017.

Thiery, Peter (2016): Aufbruch und Ratlosigkeit - BTI-Regionalbericht Lateinamerika und Karibik 2016. Bertelsmann Stiftung (Ed.). Available online at https://www.bti-project.org/fileadmin/files/BTI/Downloads/Reports/2016/pdf_regional/BTI_2016_Regionalbericht_LAC.pdf, checked on 12/7/2018.

Timmerman, Peter (1981): Vulnerability, Resilience and the Collapse of Society: A Review of Models and Possible Climatic Applications. Toronto, Canada: Institute for Environmental Studies, University of Toronto.

Treib, Oliver; Bähr, Holger; Falkner, Gerda (2007): Modes of governance. Towards a conceptual clarification. In: *Journal of European Public Policy* 14 (1), 1–20. DOI: 10.1080/135017606061071406.

UN (2018): Goal 6: Ensure access to water and sanitation for all. Edited by United Nations. Available online at https://www.un.org/sustainabledevelopment/water-and-sanitation/, checked on 12/7/2018.

UNCED: United Nations Conference on Environment and Development (1992): Principle No. 4 - Water has an economic value in all its competing uses and should be recognized as an economic good. In: The Dublin Statement On Water And Sustainable Development. Available online at.: https://www.wmo.int/pages/prog/hwrp /documents/english/icwedece.html#p4, checked on 10/02/2014.

UNDP: United Nations Development Program (2014): What is water governance? Online: http://www.watergovernance.org/sa/node.asp?node=846, checked on 10/02/2014.

UNESCO (2006): Water a shared responsibility. Paris, New York: Unesco Publ; Berghahn Books (The United Nations world water development report / United Nations 2). Available online at http://www.unesco.org/ulis/cgi-bin/ulis.pl?catno=145405.

UNESCO (2015): Water for a sustainable world. Paris: UNESCO (The United Nations world water development report, 6.2015). Available online at http://www.unesco.org/ulis/cgi-bin/ulis.pl?catno=231823.

UNESCO (2017): Wastewater. The untapped resource. Paris: UNESCO (The United Nations world water development report, 2017). Available online at http://www.unesco.org/ulis/cgi-bin/ulis.pl?catno=247153.

Universidad de Playa Ancha (UPLA) (2015): Municipalidad de Petorca inauguró Oficina de Asuntos Hídricos. Available online at http://www.upla.cl/noticias /2016/06/14/municipalidad-de-petorca-inauguro-oficina-de-asuntos-hidricos/, checked on 12/8/2018.

UPLA (2014): Informe Del Proyecto. Sistemas Participativos De Gestion Del Agua Y Desarrollo Socio-Economico Sostenible De La Cuenca Del Río Petorca. PN-2014-8. Edited by Universidad de Playa Ancha. Valparaíso.

Valdés-Pineda, Rodrigo; Pizarro, Roberto; García-Chevesich, Pablo; Valdés, Juan B.; Olivares, Claudio; Vera, Mauricio et al. (2014): Water governance in Chile. Availability, management and climate change. In: *Journal of Hydrology* 519, 2538–2567. DOI: 10.1016/j.jhydrol.2014.04.016.

Vennesson, Pascal (2008): Case studies and process tracing: theories and practices. In: Della Porta, Donatella; Keating, Michael (Eds.): Approaches and Methodologies in the Social Sciences. A Pluralist Perspective. Leiden: Cambridge University Press, 223–239.

Vergara Blanco, Alejandro (2014): Crisis institucional del agua. Descripción del modelo jurídico, crítica a la burocracia y necesidad de tribunales especiales. Santiago, Chile: Legal Publiahing Chile.

Walby, Sylvia (2007): Complexity Theory, Systems Theory, and Multiple Intersecting Social Inequalities. In: *Philosophy of the Social Sciences* 37 (4), 449–470. DOI: 10.1177/0048393107307663.

WBGU (2011): Factsheet No. 3/2011. Global Megatrends. Wissenschaftlicher Beirat der Bundesregierung Globale Umweltveränderungen. Berlin. Available online at: https://www.wbgu.de/fileadmin/user_upload/wbgu/publikationen/factsheets/fs3_20 11/wbgu_fs3_2011_en.pdf, checked on 30/7/2019.

Whaley, Luke; Cleaver, Frances (2017): Can 'functionality' save the community management model of rural water supply? In *Water Resources and Rural Development* 9, 56–66. DOI: 10.1016/j.wrr.2017.04.001.

WHO (2003): Boron in Drinking-water. ackground document for development of WHO Guidelines for Drinking-water Quality. Available online at https://www.who .int/water_sanitation_health/dwq/boron.pdf, checked on 12/7/2018.

WHO (2005): Mercury in Drinking-water. Background document for development of WHO Guidelines for Drinking-water Quality. World Health Organization. Available online at: https://www.who.int/water_sanitation_health/dwq/chemicals/mercury final.pdf, checked on 15/6/2020.

WHO; UNICEF (2017): Progress on drinking water, sanitation and hygiene. 2017 update and SDG baselines. Geneva: World Health Organization and the United Nations Children's Fund.

World Economic Forum (2018): The Global Risks Report 2018. 13th Edition. Cologny/Geneva Switzerland. Available online at http://www3.weforum.org/ docs/WEF_GRR18_Report.pdf.

Wuppertal Institut (Ed.) (2006): Fair Future. Ein Report des Wuppertal Instituts. Begrenzte Ressourcen und globale Gerechtigkeit. Verlag C. H. Beck oHG: München.

Zarricueta Carmona, Paula (2015): Obras y Proyectos de Infraestructura de Riego en la Provincia de Petorca. Edited by Ministerio de Obras Públicas. Available online at https://www.camara.cl/pdf.aspx?prmID=49493&prmTIPO=DOCUMENTOCOMIS ION, checked on 12/8/2018.